D0936212

MICROSCOPY HANDBOOKS 39

Introduction to Scanning Transmission Electron Microscopy

Introduction to Scanning Transmission Electron Microscopy

Robert J. Keyse
Materials Science and Engineering,
University of Liverpool,
Liverpool, UK

Anthony J. Garratt-Reed
Massachusetts Institute of Technology,
Cambridge, Massachusetts, USA

Peter J. Goodhew
Materials Science and Engineering,
University of Liverpool,
Liverpool, UK

Gordon W. Lorimer
Materials Science Centre,
The University of Manchester and UMIST,
Manchester, UK

Springer

In association with the Royal Microscopical Society

Robert J. Keyse, Anthony J. Garratt-Reed, Peter J. Goodhew and Gordon W. Lorimer
respectively Materials Science and Engineering, University of Liverpool, Liverpool, UK; MIT, Cambridge, Massachusetts, USA; Materials Science and Engineering, University of Liverpool, Liverpool, UK; Materials Science Centre, The University of Manchester and UMIST, Manchester, UK

Published in the United States of America, its dependent territories and Canada by arrangement with BIOS Scientific Publishers Ltd, 9 Newtec Place, Magdalen Road, Oxford OX4 1RE, UK

First published 1998

A CIP catalog record for this book is available from the British Library.

Library of Congress Cataloging-in-Publication Data
An introduction to STEM / Robert J. Keyse . . . [et al.].
p. cm. -- (Microscopy handbooks)
Includes bibliographical references (p.) and index.
ISBN 0-387-91517-6 (softcover : alk. paper)
1. Scanning transmission electron microscopy. I. Keyse, Robert J. II. Series: Microscopy handbooks (Springer-Verlag New York Inc.)
QH212.S34I58 1998
802'.8'25--cd21 97-31937
CIP

ISBN 0 387 91517 6 Springer-Verlag New York Berlin Heidelberg SPIN 19900598

Springer-Verlag New York Inc.
175 Fifth Avenue, New York,
NY 10010-7858, USA

Production Editor: Rachel Offord.
Typeset by Poole Typesetting (Wessex) Ltd, Bournemouth, UK.
Printed by Biddles Ltd, Guildford, UK.

Front cover: A set of images of magnesium oxide cubes taken with different STEM detectors (see *Figure 4.9*).

Contents

Abbreviations ix

Preface xi

1. Why STEM?—STEM versus TEM 1

Introduction 1
Image formation 4
Beam convergence 4
Diffraction patterns 4
Instrument geometry 5
Applications of STEM 5
High resolution analysis 5
High resolution imaging 6
Analysis of biological macromolecules 8
How to use this book 9
References 9

2. STEM optics 11

Introduction 11
Field emission gun (FEG) 12
Gun lens (GL) 14
Gun alignment 15
Differential pumping apertures (DPAs) 15
Virtual objective aperture (VOA) 15
Beam blanking 16
Pre-specimen optics 16
Condenser alignment and stigmator 16
Condenser lenses (C1, C2) 17
Scan coils 18
Objective alignment 18
Selected area diffraction (SAD) aperture 19
Objective lens region 19
Objective/condenser stigmator 19
Objective lens (OL) 20
Objective aperture 21

Specimen holder 21
Post-specimen optics and detectors 22
Secondary electron detector 22
Post-specimen alignment coils 23
Post-specimen lenses 23
Annular dark field (ADF) detector 23
Diffraction pattern observation screen (DPOS) 24
Collector aperture 24
Bright field (BF) detector 25
X-ray and energy loss spectrometers 25
Summary 26

3. The specimen 27

Geometry 27
Introduction 27
X-ray detectors 27
Specimen preparation 29
Tilting possibilities 30
Airlock 33
Contamination 34
Beam damage 35

4. Imaging in the STEM 37

Introduction 37
General discussion of probe forming 38
A few numbers and formulae (facts and figures) 39
Processes in image formation 41
STEM detectors 41
Some typical images from a STEM 43
Bright field STEM images 43
Annular dark field STEM images 43
Secondary electron and Auger images 46
Resolution 47
Introduction 47
Defining resolution 47
Limits to resolution 48
Scattering inside specimens 50
Atomic number contrast 50
Thickness effects 50
Comparisons with CTEM 51
Relationships with diffraction 52
A few more numbers 53
References 54

5. **Diffraction in the STEM** 55

Introduction 55
Lens effects 55
Reciprocity 56
Selected area diffraction 57
TEM image magnification 57
Selected area diffraction in TEM 58
Selected area diffraction in STEM 58
Post-specimen compression 61
Other types of diffraction pattern 62
Convergent beam electron diffraction (CBED) 62
Microdiffraction 64
High resolution STEM imaging 65
Limits to diffraction 67
Summary 68

6. **Microanalysis in the STEM** 69

Introduction 69
Energy dispersive X-ray microanalysis in the STEM 70
EDXS detectors 70
X-ray detector windows 71
Windowless detectors 72
EDXS spectrum details 72
Escape peaks 74
Sum peaks 75
Coherent bremsstrahlung (CB) 75
Quantitative X-ray microanalysis 75
Cliff–Lorimer thin film method 75
Hall method 76
X-ray absorption 76
Limits of EDXS analysis 77
Light element analysis 79
Energy resolution 79
Dead-time 80
Electron energy-loss spectroscopy 80
Energy loss spectrometers 81
Interfacing to the microscope 82
Data collection systems 83
Energy transitions 84
Details of the energy loss spectrum 86
Near-edge structure 87
Extended fine structure 88
Specimen thickness effects 89
Detection limits 89
Analytical strategy 90
Conclusion 90

7. Mapping in the STEM **91**

Introduction 91
X-ray mapping (including linescans) 91
Digital mapping 92
Digital linescans 94
High-angle annular dark-field imaging (HAADFI) 95

8. Limits to STEM and advanced STEM **97**

Limits to microprobe analysis 97
Developments of the FEG 98
Spherical aberration 99
Operation of the STEM 100
Spectrum imaging 100
Beam damage and drilling holes 101

Appendices **105**

Appendix A: Glossary 105
Appendix B: Further reading 109

Index **111**

Abbreviations

ADF	annular dark field
AEM	analytical electron microscope
BF	bright field
BS	backscattered
CB	coherent bremsstrahlung
CBED	convergent beam electron diffraction
CCD	charge-coupled device
CRT	cathode ray tube
CTEM	conventional TEM
DPA	differential pumping aperture
DPOS	diffraction pattern observation screen
EBIC	electron-beam induced conductivity
EDXS	energy dispersive X-ray spectrometer
EELS	electron energy-loss spectrometer/spectroscopy
EHT	extra-high tension
ELNES	energy-loss near-edge structure
EMMA	electron microscope microprobe analyser
EXELFS	extended energy-loss fine structure
FE	field emission
FEG	field emission gun
FEG-STEM	field emission gun STEM
FWHM	full width at half maximum
GIV	gun isolation valve
GL	gun lens
HAADF(I)	high-angle ADF (imaging)
HREM	high resolution electron microscopy
i.d.	inner diameter
o.d.	outer diameter
PEELS	parallel EELS
PMT	photomultiplier tube
SAD	selected area diffraction
SE	secondary electron
SEM	scanning electron microscop(e/y)
STEM	scanning transmission electron microscop(e/y)
(S)TEM	TEM with scanning facility (STEM in a conventional TEM)

TEM	transmission electron microscop(e/y)
TOA	take-off angle
UHV	ultra-high vacuum
VOA	virtual objective aperture
WDXS	wavelength dispersive X-ray spectrometer
YAG	yttrium–aluminium–garnet

Preface

1997 was the 'Year of the Electron' because it marked the centenary of the celebrated discovery of the smallest of the fundamental particles that make up ordinary matter, and which has proved to have so many remarkable properties that it has become, after light, the most widely used of the particles in scientific and technological applications. The excitement generated during this century by the wonderful electron is not showing any sign of diminishing as we approach the millennium. In virtually all walks of life people are making use of electrons, while in scientific research new ways to explore the world by using electrons are still being developed.

Electron microscopy and microanalysis have a long tradition in technological innovation. Electron diffraction and electrodynamics represent important experimental and theoretical developments in 20th century physics. There is a wealth of work still to be carried out to discover further secrets in the realms of science and technology; and surely it is a safe bet that electrons will continue to play a pivotal role in the years ahead. Where else can one find the special theory of relativity and quantum mechanics so well integrated, and the apparently conflicting wave and particle descriptions of matter so clearly illustrated than in a high-voltage field emission electron microscope - an instrument that is frequently described (and designed!) in terms of classical geometrical optics?

This book grew out of a perceived need for graduate students to find a source of answers to the type of question: 'I have used (or I am going to use) STEM in my research, where can I find out more about the techniques and the instrumentation?' It seemed to the authors that here was a worthy subject for a Royal Microscopical Society Handbook and so this book was conceived. It attempts to inform students and other users of the possibilities and practicalities of the application of STEM techniques to the examination of materials.

Robert J. Keyse
Anthony J. Garratt-Reed
Peter J. Goodhew
Gordon W. Lorimer

Acknowledgements

The authors would like to express their gratitude to colleagues who have supplied examples of their work, the students who asked all those questions and to the production staff at BIOS for their patience. Individual thanks are due to Peter Kenway for reviewing the manuscript and making many helpful comments, Graham Cliff for his enthusiasm, Rob Devenish for his expertise, Helen Davock for helpful comments and finally our families for their support.

1 Why STEM?—STEM versus TEM

1.1 Introduction

What is STEM? The acronym literally means scanning transmission electron microscopy (or microscope) and implies transmission electron microscopy (TEM) performed with a scanned, focused electron beam. Since conventional TEMs have been available for about 40 years and have been developed to a state at which they have an extensive range of capabilities, we need to answer the question 'why bother with STEM?'. We aim to answer this question in this first chapter, before going on to describe the operation and some of the theory of the STEM in later chapters.

In principle, STEM is a very straightforward technique, as *Figure 1.1* shows: a fine electron probe is scanned across a thin specimen and the intensity of the transmitted electron signal is measured using one or more electron detectors. An image is thus built up point by point, just as in a television or a conventional scanning electron microscope (SEM). If other detectors such as an X-ray detector or an electron energy-loss spectrometer are attached then chemical microanalysis can be performed point by point. If a secondary electron detector is incorporated then an SEM image can also be obtained.

You might argue that all this can be done either in a conventional SEM or in a conventional TEM modified to allow scanning of the beam. The essential difference between these conventional microscopes and a true STEM instrument lies in the (small) size of the probing electron beam and the current density (electrons per second per unit area) it can deliver to the specimen.

A TEM may have a thermionic electron source with a tungsten or lanthanum hexaboride (LaB_6) emitter, or a field emission (FE) source. A TEM with the appropriate detectors, scanning coils, and any of these sources can act as a STEM. However, the technique comes into its own when used with a high brightness FE source because only then can a very fine probe (less than 1 nm in diameter) be used to deliver a high

(a)

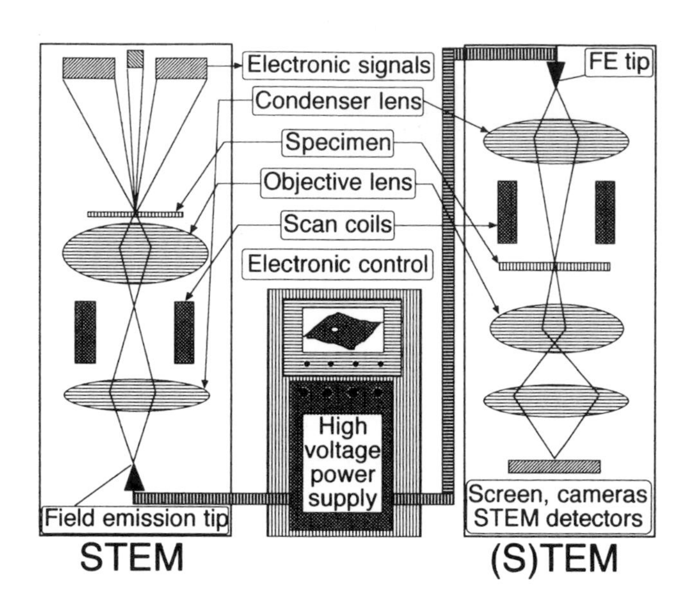

(b)

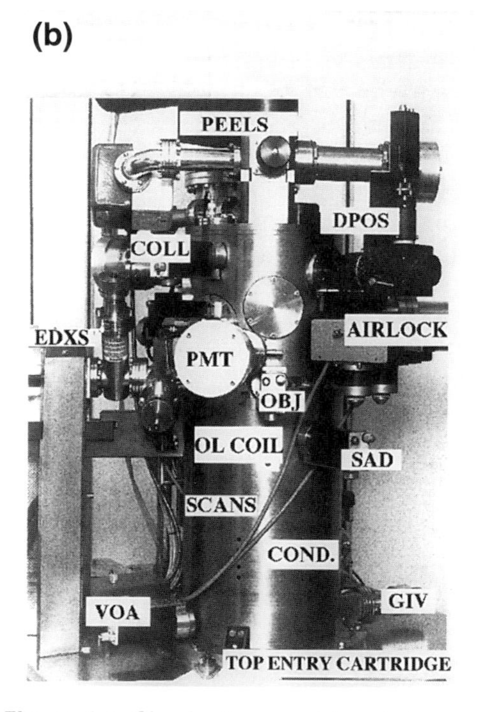

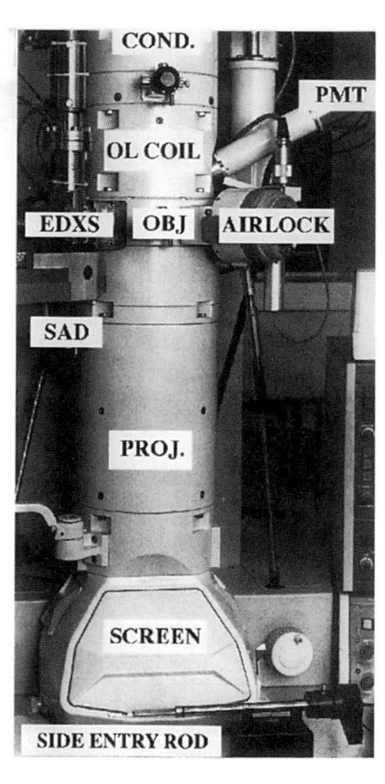

Figure 1.1. Simple diagrams (a) and photographs (b) of a STEM and (S)TEM side by side, in their usual configurations. The electron beam goes from bottom to top in a STEM but from top to bottom in a (S)TEM. PEELS, parallel electron energy-loss spectrometer; COLL, collector aperture; DPOS, diffraction pattern observation screen; PMT, photomultiplier tube; EDXS, energy dispersive X-ray spectrometer; OBJ, objective aperture; OL COIL, objective lens coil; SAD, selected area diffraction aperture; COND., condenser lenses; VOA, virtual objective aperture; gun isolation valve; PROJ., projector lenses.

current (more than 0.5 nA, or about 10^{10} electrons per second). If chemical analysis or diffraction is to be carried out with the fine probing beam stationary for many seconds on a single part of the specimen, then a clean specimen and a clean high vacuum environment are essential, or the part of the specimen being examined will rapidly be contaminated. When a FE source is combined with an ultra-high vacuum (UHV) scanning transmission electron microscope column the instrument is usually known as a 'dedicated' STEM.

STEM instruments tend to cost more than TEM instruments because there are UHV requirements, additional components in the microscope, and extra electronics to drive them. One advantage of STEM is the ease of interfacing to a computer, since by its very nature a STEM is an electronic microscope. High direct magnification TEM images are sometimes only dimly visible on the phosphor screen and they often have only limited contrast; in STEM the image stays bright even at the highest magnifications and contrast can be adjusted electronically. It is usual to conduct TEM experiments in a darkened room (so your eyes become dark adapted), but with a STEM one normally only dims the lights for comfort. A lot of the guesswork is taken out of STEM investigations, for example when focused at high magnification the image remains focused at any lower magnification. Also, since only those parts of the specimen that are scanned are irradiated, there is no hidden beam damage occurring outside the field of view. At first sight STEM may appear to be a complicated and sophisticated technique, but times are changing and ease of use has become a major design goal.

It may be useful at this point to clarify some of the acronyms which are used in discussing this area of microscopy. We give some of the most common in *Table 1.1*.

Although both TEM and STEM instruments produce transmission electron images there are a number of very significant differences between the imaging modes. Let us consider four of these differences immediately.

Table 1.1. Some commonly used acronyms

Acronym	Meaning
AEM	Analytical electron microscope (implicitly an analytical TEM)
CTEM	Conventional TEM
EDXS	Energy dispersive X-ray spectrometer (often EDX or EDS)
EELS	Electron energy-loss spectrometer (or Spectroscopy)
FEG-STEM	Field emission gun STEM
HREM	High resolution electron microscopy (usually in a CTEM)
SEM	Scanning electron microscope
STEM	SEM with transmission facility (usually a dedicated STEM)
(S)TEM	TEM with scanning facility (STEM in a conventional TEM)
TEM	Transmission electron microscope

1.2 Image formation

A conventional TEM (CTEM) produces a parallel image on a screen, photographic plate, or camera, in which all pixels are recorded or observed simultaneously and the image magnification is controlled by the projector lenses after the beam has passed through the specimen. In a STEM the image is collected in series, pixel by pixel, and no lenses are required for image magnification. In principle this helps to preserve a high quality image in STEM because aberrations in 'post-specimen' lenses do not influence the image quality. However, in practice the STEM probe-forming lens aberrations prove to be a limitation and the ultimate performance of a STEM is usually aberration-limited just like a TEM. Providing the aberrations are limited the ultimate beam diameter can be made very small; it is possible to use a STEM for high resolution electron microscopy (HREM) as shown in Chapter 4 (see *Figure 4.8*).

1.3 Beam convergence

When a beam of electrons is focused to form a very fine electron probe it is almost inevitable that it will be highly convergent (in the TEM one often converges the beam to a spot at the specimen). Consequently, at all times during STEM image formation and microanalysis the beam is very convergent. In contrast, images are usually obtained with a TEM by using a 'defocused' beam of almost-parallel electrons. We will show in Chapter 4 that this affects the way in which we can use a STEM to form diffraction contrast images and it also affects the appearance of diffraction patterns, as we will show in Chapter 5. We shall invoke the *principle of reciprocity* to explain these effects; that is, that any optical system behaves identically if the direction of the radiation (light or electrons) is reversed.

1.4 Diffraction patterns

The highly convergent probe in a STEM emerges from the far side of the specimen as a convergent beam diffraction pattern. It can be seen from *Figure 1.1a* that at all times there is only ever a diffraction pattern in the column of a STEM. Consequently, if we stop the beam scanning and record the distribution of electron paths in the column of the microscope, we have recorded the diffraction pattern from a small region of the specimen. Many dedicated STEMs are not designed to do this with the greatest efficiency, and therefore are not primarily used as diffraction instruments, but the potential is clearly there, as we will show in Chapter 5 (see *Figure 5.5*).

1.5 Instrument geometry

There is no need for projector lenses after the specimen in a STEM, although as shown in Chapters 2 and 5 there are some advantages to be gained from post-specimen lenses. Potentially there is a great deal of space after the specimen in a STEM which can be used to house a multitude of electron detectors. This is in stark contrast to a CTEM, in which there is only very restricted space between the specimen and the subsequent lenses. In a CTEM the major part of the objective lens action is after the specimen, while in a STEM it is before the specimen. Dedicated STEMs often have an inverted column with the electron source near the floor and the electrons travelling upwards. This means that the space after the specimen is at the top of the column and is easily accessible. This space may contain several electron detectors (e.g. bright field, dark field, and secondary electron detectors; see Chapters 2, 4, and 7) and an electron energy-loss spectrometer (EELS) system (see Section 6.6). Additional possibilities include one or two X ray detectors (see Section 6.2), backscattered electron detectors, Auger electron spectrometers, and light sensitive detectors, though these are usually before the specimen.

1.6 Applications of STEM

Before discussing the microscope itself we will look at three very different applications of STEM in three very different areas of science.

1.6.1 High resolution analysis

Figure 1.2 shows the results of an analysis carried out along a line across the boundary between a particle of titanium carbide (TiC) and the intermetallic titanium aluminide ($TiAl_3$) in which it is partially embedded (Worsley, 1997). The beam of the STEM was moved to each of 32 points along a line only 40 nm long and the four curves show the variation of the characteristic X-ray signal from C, Al, Si and Ti along this line.

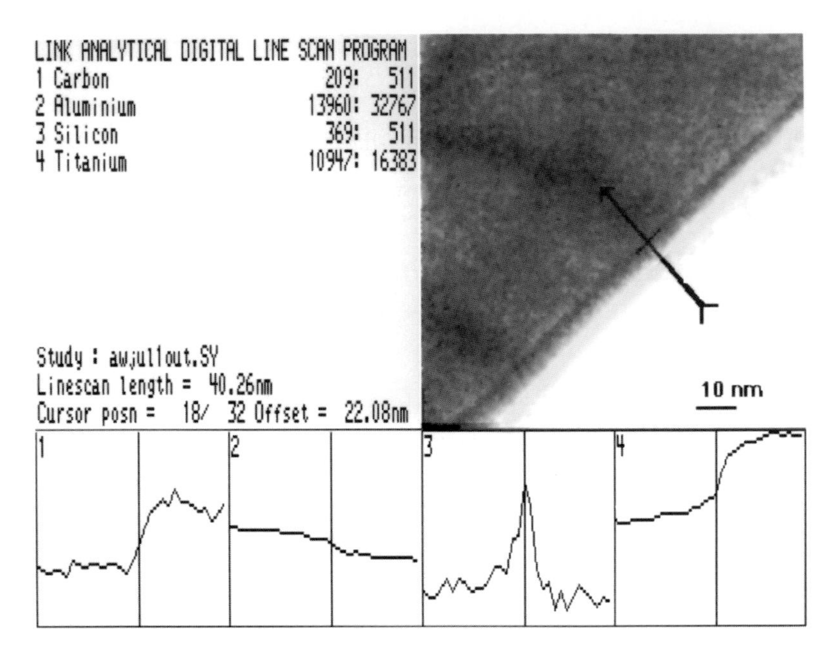

Figure 1.2. A digitally acquired EDXS linescan across a phase boundary in an aluminium master-alloy. For details please see text. (Courtesy of Dr G. Tatlock, University of Liverpool.)

The intensity of the signals from C and Ti (curves 1 and 4 in *Figure 1.2*) rise at the interface, as the electron beam enters the carbide particle (the C signal rises above its background). The aluminium signal (curve 2) drops at the same point, but not completely since the electron beam is passing through some $TiAl_3$ above (or below) the TiC. The most interesting signal comes from silicon (curve 3) and shows the segregation of Si to the particle/matrix boundary. The segregant only appears in a region 5 nm wide and more detailed analysis shows that it comes from a thin layer of silicon atoms at the boundary.

From the graded contrast in the image we can see that the boundary does not appear to be perfectly perpendicular to the plane of the foil, so it is not quite parallel to the electron beam (see *Figure 3.4*). This means that the present example is not the best possible illustration of the resolution available in energy dispersive X-ray spectroscopy (EDXS) analysis, but it is an example of the sensitivity which is possible with the intense and narrow electron beam of a field emission gun (FEG)-STEM. (A higher spatial resolution example of an EDXS linescan obtained manually in a STEM is shown in *Figure 7.2*.)

1.6.2 High resolution imaging

Figure 1.3a shows a FEG-STEM image of a self-organized array of gold particles (Fink *et al.*, 1997). The image was taken with a VG601UX STEM using a beam diameter of 1 nm and it shows that the particles, of

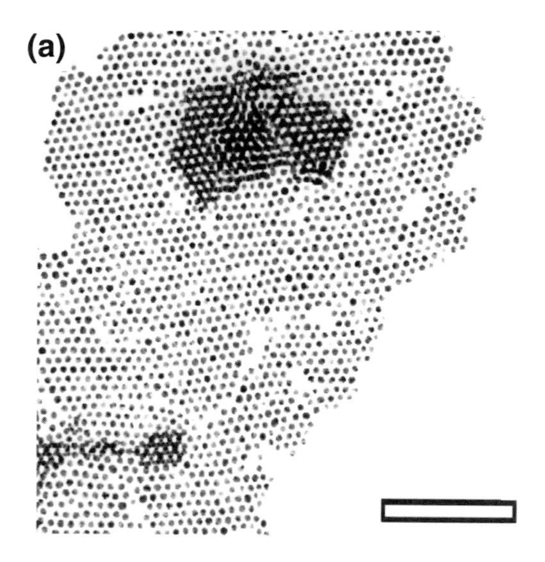

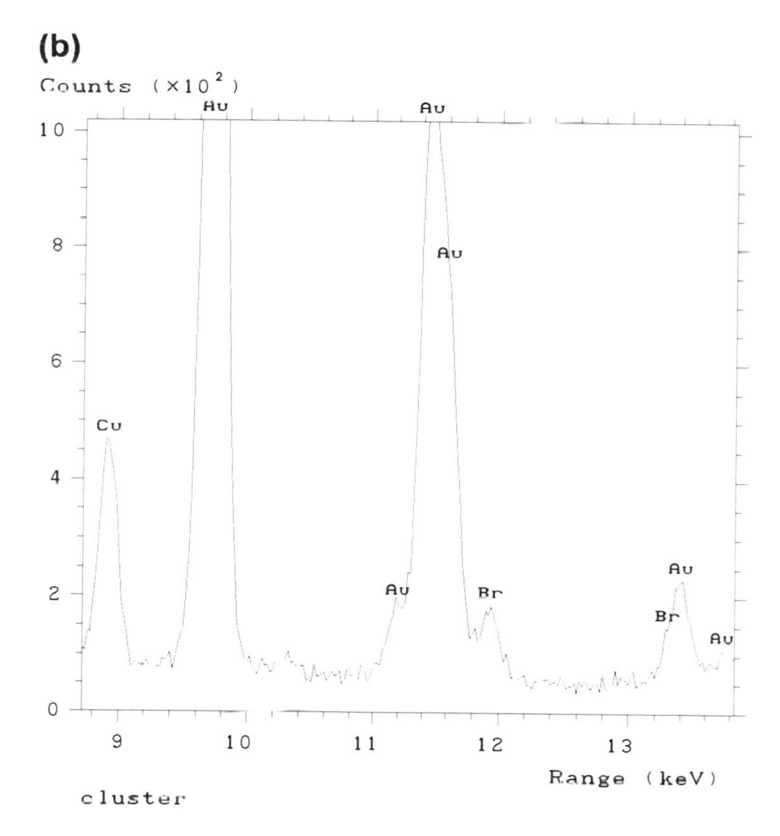

Figure 1.3. (a) A bright field STEM image of rafts of gold islands. Note carefully the way the layers are packed together, indicating that a repulsive force may exist between the particles. Bar=100 nm. (Courtesy of Dr C. Kiely, University of Liverpool.) (b) An EDXS spectrum from a group of gold islands showing bromine, thought to be concentrated between the particles.

2 STEM optics

2.1 Introduction

This chapter describes the features of a FEG-STEM instrument in detail; most of the discussion is general enough to be of use with instruments of several types. In particular we concentrate on describing the electron optics in terminology specific to the STEM mode of operation. It is not sufficient to say that the only difference between a dedicated STEM and a TEM (with a scanning attachment) is the existence of post-specimen lenses. A dedicated STEM is more closely related to an SEM than a TEM in its optical design, so it might be described as an SEM with a transmission attachment or a S(T)EM. In the immediate future most new analytical electron microscopes (AEMs) are likely to be TEMs with scanning attachments or (S)TEMs. They will probably be based on a TEM in design, but designed to optimize the formation and scanning of a very fine electron beam (see Chapter 8 for more on the future).

Figure 2.1 below and the following text are closely associated; refer to the diagrams and use the text as a commentary throughout the book. Two instrument configurations are shown: a typical STEM with its electron gun at the bottom (on the left of the diagram) and a typical (S)TEM with the electron gun at the top. The latter arrangement is more familiar to most TEM users but the 'gun on the floor' geometry was popular for many years with the instrument manufacturer VG (latterly Fisons). We will consider the essential components of STEM instruments starting with the electron gun and finishing with the image collection system. This is top-to-bottom for a (S)TEM but bottom-to-top for a VG STEM.

The key components, in order, are the electron gun (the source of electrons), the gun alignment controls, apertures to limit the diameter of the electron beam, condenser lenses to collimate it, scan coils to move it, objective lens controls to form the probe, the specimen holder, and finally an array of detectors to pick up the signals from the specimen.

The probe size, the convergence angle, the current in the probe, and the electron energy are four parameters which determine the nature and quality of the image or spectrum obtained from the specimen by the instrument.

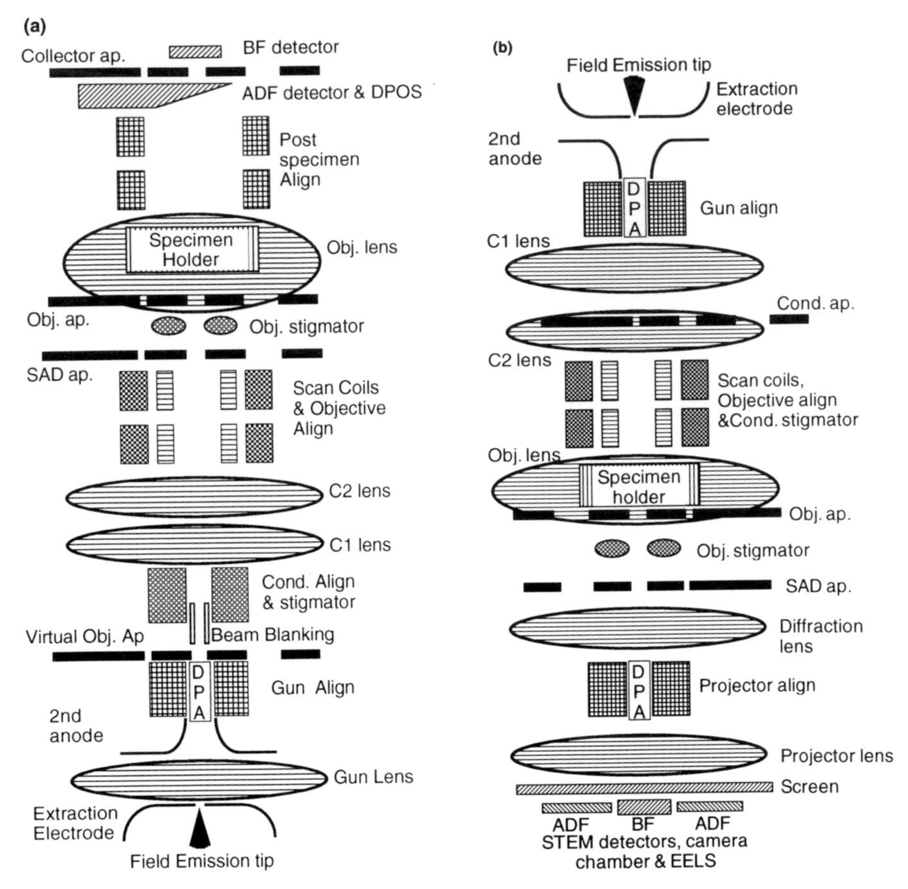

Figure 2.1. Schematic diagrams of two systems side-by-side: (a) STEM, (b) (S)TEM. ADF, annular dark field; BF, bright field; DPA, differential pumping aperture; DPOS, diffraction pattern observation screen; SAD, selected area diffraction.

2.2 Field emission gun (FEG)

A field emission gun (FEG) (see *Figure 2.2*) consists, in essence, of a metal surface (the 'tip') with a radius of curvature of about 50 nm separated from a plane metal surface by a millimetre or so. The plane surface (extraction electrode) has a small hole drilled in it, and a potential difference is sustained between the surfaces of a few thousand volts. At the surface of the tip the electric field strength may reach an intensity of around 10^{11} V m^{-1} and interesting phenomena may then be observed.

Electrons inside the tip normally stay within the confines of the surface because their wave amplitude falls rapidly with distance inside the barrier. They are thus unable to penetrate the surface potential. However, the intense external electric field makes the barrier appear to become very narrow and some electrons seem to 'tunnel' through and escape. The small current of electrons can be increased if more kinetic

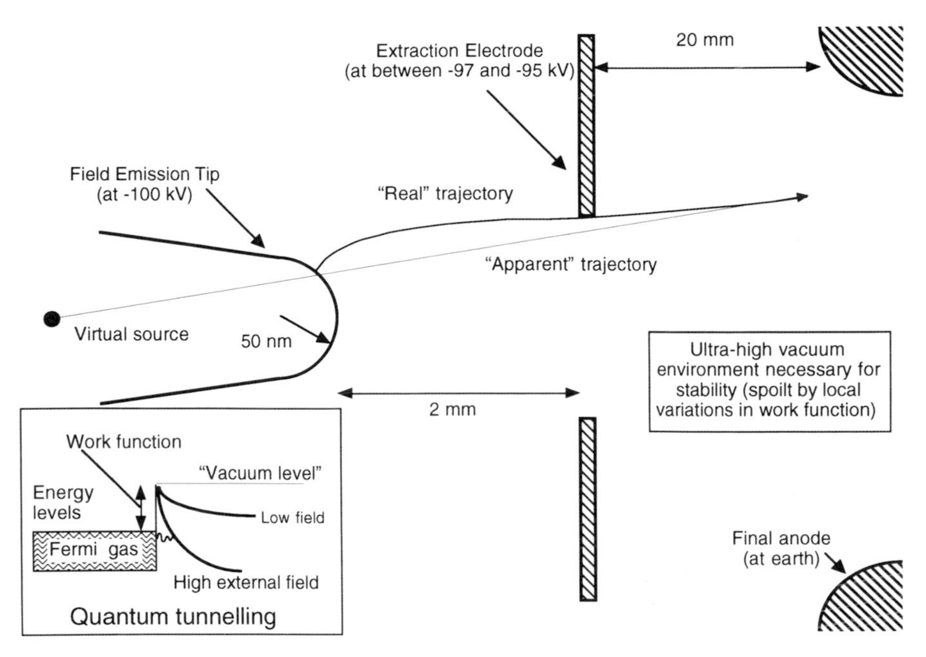

Figure 2.2. Schematic diagram of a simple 'triode' field emission gun. The inset shows a representation of an energy level diagram showing the quantum tunnelling process.

energy is available, since this allows them to reach parts of the barrier that are even thinner and more will get away. Give them too much energy, however, and they'll find it easier to cross the barrier by more conventional (thermionic) means.

The electron source in an FEG is usually made out of a single tungsten crystal, sharpened electrolytically. This offers a huge increase in brightness for a given accelerating voltage over the conventional tungsten hairpin thermionic source. In many designs the improvement in brightness (measured in amps per cm^2 per steradian) is as much as a factor of 1000. This means that an FEG is usually the gun chosen for a dedicated STEM, where it is essential that a high electron current is focused into a very fine beam.

The small effective source size of a field emission tip, that is, the size of the virtual source, where electrons seem to be coming from, permits the use of a small overall demagnification factor. Typically the source size is of order 5 nm in diameter and a demagnification of about 10 is sufficient to produce a probe size of 0.5 nm. The probe-forming lens system does not need to be very powerful. The high brightness of the field emission source allows a sufficiently large current (several μA) to be emitted from the (small) source with a small angle of divergence. Thus a final probe current of the order of nA, which is useful analytically, can be produced in a probe of a size measured in nm, which is useful microscopically. The average current density inside a probe of 1 nm, full-width half-maximum (FWHM) diameter, which contains 0.5 nA, is about 640 A mm^{-2}. Electron probes have soft edges; they are not 'top-hat' shaped in profile, most often they are more like a Gaussian curve (see Chapter 4 for more details).

The field emission tip can be either left at room temperature or heated. A cold field emission tip produces electrons with a small energy spread of about 0.3 eV (a natural consequence of the quantum tunnelling emission process and its low temperature). There are several FEG designs that use a warm tip (around 1500–2000 K) and the emitted electrons have a higher energy spread of about 0.6 eV. The vacuum requirement for the cold FEG is the most stringent because gas adsorbed on the tip surface tends to stick and raises the work function (the potential barrier that the electrons have to tunnel through, see *Figure 2.2*). Warm or thermal FEGs have more modest vacuum requirements than cold FEGs and consequently are cheaper to make, needing pressures of only 10^{-8} mbar rather than 10^{-11} mbar. The Schottky source consists of a tungsten tip with a thin film of zirconia, lowering the work function. This has a larger source size but a smaller angular beam spread and a larger beam current than the cold FEG source. It is a reasonably new type of source that shows promise for the future as it has several advantages over the FEG sources, most importantly in terms of emission stability.

The majority of (S)TEMs in use do not exploit the undoubted advantages of any type of FEG, and the most common source of electrons is still the heated tungsten hairpin filament, which emits thermionically at around 3000 K. The next best things to the tungsten hairpin source are the rare earth hexaboride sources (LaB_6 and CeB_6), both of which are used as heated, pointed filaments. Hexaboride sources are thermionic but operate at a lower temperature and have a smaller source size than the tungsten hairpin; they need a better vacuum than the tungsten hairpin but not as good a vacuum as the FEG. Many (S)TEMs have vacuum systems capable of LaB_6 operation, which is likely to remain the source of choice for most laboratories doing routine analytical work. Many of the newest (S)TEMs have baked, ion-pumped guns of sufficiently low pressure to operate thermally assisted FEGs.

Table 2.1 summarizes some of the parameters of each type of source, although in the remainder of what follows we concentrate on the FEG source.

2.2.1 Gun lens (GL)

To further collimate the beam and reduce chromatic aberrations in the FEG most microscopes are fitted with a gun lens (GL). The best lenses have a magnetic field with a maximum flux density located in the gap between the extraction electrode and the final anode (see *Figure 2.2*), the region where the electrons accelerate through a high potential difference. High voltage microscopes (200–400 kV) in general tend to have long acceleration tubes with electrostatic focusing elements between the extraction electrode and the accelerating region.

Table 2.1. Summary of source types

Source	Brightness (A cm^{-2} sr^{-1})	Temperature (K)	Energy spread (eV)	Vacuum (mbar)
Cold FEG	10^9	300	0.3	10^{-11}
Thermal FEG	10^8	2000	0.5	10^{-9}
LaB_6	10^6	1500	1.0	10^{-7}
Tungsten	10^5	3000	2.0	10^{-6}

Without a specific GL the electrostatic acceleration itself acts as a weak lens (electron trajectories tending to become more parallel). If a large vacuum pressure difference must be sustained between the gun and optical column chambers (e.g. as in the cold-FEG case) then an electron beam cross-over is often needed near to a small differential pumping aperture (DPA, see Section 2.2.3). The cross-over is produced by the action of the gun lens.

2.2.2 *Gun alignment*

The GL is electrically or mechanically tilted to maximize the beam current emerging from the gun via the DPAs. It also requires some electrical alignment to maintain the apparent source position stable with respect to voltage fluctuations on the tip. To achieve this alignment coils are placed after the GL, as shown in *Figure 2.1.*

2.2.3 *Differential pumping apertures (DPAs)*

To maintain a stable emission current over time a cold FEG needs an ultra-low pressure or, as is more usually stated, an ultra-high vacuum (UHV) environment. To facilitate the conflicting requirements of rapid sample exchange (and possibly slow recovery of specimen chamber pressure) and the necessary UHV environment for field emission, fixed apertures are placed in the beam path. Ideally there is a very low leak rate of gas molecules from the main column to the gun chamber, while the electron beam passes unimpeded in the opposite direction.

The DPAs are beam limiting and can be seen as fixed first condenser apertures similar to those that are common in many conventional high voltage (200–400 kV) electron microscopes. The GL is normally adjusted to form a focused beam at a point before the DPA, although it can be placed after the aperture for high current applications (e.g. Auger microscopy). Typically the DPA consists of two apertures about 3 mm apart, with the one nearest the tip somewhat larger than the other (both being fractions of a millimetre in diameter).

2.2.4 *Virtual objective aperture (VOA)*

Just after the DPA is a moveable aperture mechanism used primarily during X-ray analysis. It is a beam defining aperture, that is, only those electrons that pass through the aperture go on to reach the sample. In the VG STEM it is located before the condenser lenses. When used under particular condenser lens conditions the VOA and objective aperture planes are conjugate and therefore the VOA acts, electron optically, as a replacement for the standard objective aperture. Its position is governed solely by the requirement that it is beam defining, is out of sight of the X-ray detector, and is located far from the sample. It can be seen (functionally) as a condenser aperture. Most (S)TEMs use the condenser aperture for probe definition (see Section 2.4.3).

2.2.5 Beam blanking

Two beam blanking plates may be fitted inside the vacuum after the VOA. They may be connected to a high-speed/high-voltage power supply that is linked to the timebase fly-back pulse (or some other input) and delivers its output to the plates, typically 1 kV, in microseconds. When energized, these plates deflect the beam so that it hits a non-beam defining aperture further along the microscope column, thus blanking the beam from the specimen. The uses for beam blanking include reduction of beam-induced damage and contamination (by allowing the beam to hit the specimen only for the minimum time needed), optimized X-ray counting at high count rates, and coincidence counting.

2.3 Pre-specimen optics

Electrons emerging from the gun enter the optical column which comprises several more lenses, apertures, and alignments. In the STEM there are additional features that facilitate the scanning of the focused probe over the specimen.

2.3.1 Condenser alignment and stigmator

When two condenser lenses (C1 and C2) are used, electrical alignment is required to align the C1 and C2 lens axes. The alignment is designed to make the probe position independent of variations in the current to the condenser lenses.

The stigmator is used primarily to obtain a symmetrically shaped probe after the condenser lens (at the selected area diffraction, or SAD aperture, see Section 2.3.5) and is often called the condenser (but sometimes the gun) stigmator. When a probe formed by C2 (see *Figure 2.3*) in the STEM is scanned over the SAD aperture, which comes after the scan coils, astigmatism may be detected in an image of the aperture and hence may be corrected with the stigmator.

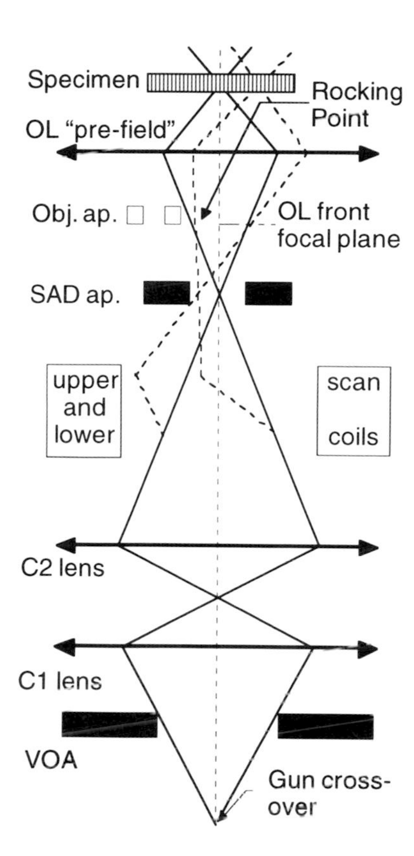

Figure 2.3. Imaging in STEM: a ray diagram for the VG STEM HB5 (HB6) series with two condenser lenses and a VOA (with C2 focused on the SAD aperture). Note that the source position is shown at the bottom and may be taken as that produced near the DPA by the GL. As shown, C1 is demagnifying and C2 is magnifying, but the diagram is not to scale. Scan coils are positioned and energized to deflect the beam in order to fill the SAD aperture, there is a cross-over at the SAD aperture for both scanned (broken line) and unscanned (unbroken line) beams. At the objective lens the beams are focused on the sample. The scanned electrons are crossing unscanned electrons at a point called the 'rocking point'. This point is at the front focal plane of the objective lens in the imaging mode of operation. The real objective aperture is normally placed at the rocking point; in practice the rocking point is electronically adjusted to coincide with wherever the objective aperture happens to be.

2.3.2 Condenser lenses (C1, C2)

There are two condenser lenses following the GL that may be used singly or together. The two lenses sit close together, C2 is after C1 in the column and is closer to the sample. In the VG STEM there are mechanical shift and tilt alignments for each lens that are used to bring the condenser lenses to the optical axis defined by the fixed objective lens. The object point for C1 is the cross-over formed by the GL (or in its absence the apparent source position itself). For analytical STEM operations the C2 image position is often the SAD aperture (see Section 2.3.5). The effect of the lenses is to vary the overall system demagnification (or spot size); and this in turn allows the probe convergence angle to be varied continuously when using the VOA. External interferences (acoustic or magnetic) can enter the column anywhere but their influence in the gun region can be minimized by operating the condenser lenses in a demagnifying mode.

2.3.3 Scan coils

Scanning the electron beam across the sample gives rise to a signal modulation that, when displayed on a monitor scanned in synchronization, is equivalent to a real-time image. Scanning a smaller area of the specimen gives rise to a higher magnification image (in any scanning system the magnification is the ratio of a displayed dimension to an equivalent scanned dimension). Low voltage/high current power supplies driving low impedance coil windings produce a magnetic deflection of the electron beam. Electrostatic scanning is not normally used since high energy electrons require thousands of volts to deflect them and sufficiently stable, yet high speed, power supplies are not commonly available. The main scanning coils for the microscope are placed outside the vacuum between the condenser lenses and the specimen, and construction materials must be rigorously non-magnetic. Special care is required to shield external magnetic fields from causing unwanted interference in the STEM, especially when operating at high magnification since then the scan coils are only carrying small currents.

The coils are organized into two sets (upper and lower) and each set consists of two orthogonal pairs of coils, symmetrically placed about the axis. Thus there are X and Y, upper and lower coil pairs. The (double) deflection given to the electron beam by the coils is adjusted electronically to produce two main scanning modes, **image** (see *Figure 2.3*) and **selected area diffraction** (see *Figure 5.2*). Careful balancing of the scanning currents allows the final two-dimensional 'raster' pattern on the sample to have minimum distortion.

2.3.4 Objective alignment

Within the scan coils assembly there are further (double deflection) coils placed to provide electrical alignment of the beam entering the objective lens field. These are adjusted and set only during a major alignment and are primarily meant to compensate for any mechanical tilt of the objective lens. The method of alignment involves reversing the objective lens current (and so its magnetic field) and correcting movement of the probe position accordingly. A second use for the coils is to provide for an electrical shift of the scanned area on the sample (beam shift). Using beam shift an automatic correction of specimen drift can be made by computer control during extended X-ray mapping experiments (see Section 7.2.1).

2.3.5 *Selected area diffraction (SAD) aperture*

Just after the scan coils in the STEM there is a moveable aperture, called the selected area diffraction (SAD) aperture by analogy with the usual CTEM aperture, which is in the first image plane after the specimen (see Section 5.2.3). Optically it is the obvious location for a beam cross-over that allows the objective lens to have a fixed object position for its probe-forming action, forming a demagnified image of the source at the specimen (see *Figure 4.3*).

A very important role for the SAD aperture is in reducing the X-ray 'hole-count' at the specimen (i.e. X-ray signals collected when the beam passes through a hole in the sample). If the probe is defined by the VOA and the C2 lens focuses the beam at the SAD aperture, then the beam passes cleanly through the aperture while X-rays produced by the VOA are stopped. Thus few stray X-rays arrive at the sample, and the analysis that results from the electron beam–specimen interaction (see Chapter 6) is 'clean'. Adjusting the magnification to prevent the beam striking the aperture edge is useful when mapping (otherwise it limits the field and can excite fluorescence from the sample).

2.4 Objective lens region

The next four subsections describe the optical components that are found closest to the specimen itself (including the specimen holder). The text is by necessity becoming more specific to one type of instrument and most of the differences between instruments are to be found in the details of this region.

2.4.1 *Objective/condenser stigmator*

A stigmator is used to alter the probe shape in such a way that it is cylindrically symmetric within the interaction volume of the specimen. In the dedicated STEM an electrostatic stigmator is placed in the front bore of the objective lens close to the specimen and provides 'prior' compensation for astigmatism caused by the objective lens and the specimen itself. (S)TEM type instruments use the condenser stigmator to perform this adjustment. The effectiveness of the stigmator is seen in the image (possibly a lattice image) formed using the scanning system and the electron detectors. The most sensitive images to use for the adjustment are the secondary electron and annular dark field images (see Section 4.3.1); the bright field image behaves just like the CTEM image.

It is possible to align (or 'balance') the stigmator electronically with the objective lens axis so that minimal probe movement occurs whilst adjusting the stigmator. Locating the objective lens axis is easily done by finding the probe-defining aperture position that minimizes probe movement at the sample with respect to deliberate objective lens current variations (the process is sometimes called 'wobbling' the lens).

2.4.2 Objective lens (OL)

This is the heart of the machine, where the sample meets the probe and the interactions that yield information about the sample (and the probe) take place. There exists a region of cylindrically symmetric magnetic flux density with the sample sitting at or near its peak intensity (often about 1 tesla). The lens action is controlled by the current passing through a coil producing the above magnetic field between the pole-pieces. The gap between the pole-faces is only a few mm, but usually contains enough space for the specimen holder, an aperture blade, and a view to the X-ray detector. Sometimes (in (S)TEMs) there is enough space for a backscattered electron detector concentrically positioned around the pole facing the entrance surface of the sample (usually above the sample).

The shape of the magnetic field (its curvature and distribution in space) determines the electron optical properties of the lens and, to a large extent, defines the microscope's performance. The highest performance is obtained when the field in the pole-piece gap is most abruptly shaped with the highest peak field and narrowest 'width'. Electron lenses are really of poor quality when compared to light-optical glass lenses, because their aberrations are almost impossible to 'correct' using combinations of convex and concave lenses (as is the case in light optics). Normal (cylindrical) magnetic lenses cannot be 'concave', they always converge electrons towards the axis; quadrupole lenses can diverge electron trajectories in one plane, though they converge them in a plane at right angles.

In discussing probe forming it is usual only to consider that part of the magnetic field up to the sample position, but obviously there is more to the magnetic field after the specimen, and this is often called the 'post' field. Axial magnetic field symmetry, before and after the specimen, determines the relative importance of the pre- and post-specimen field; essentially symmetrical fields are characteristic of many (S)TEMs, while asymmetric fields are often associated with dedicated STEMs. The origin of the difference in magnetic field symmetry is mostly related to which type of specimen holder is fitted (see Section 2.4.4).

The effect of the post-field on the emerging electrons in a STEM is to cause them to reduce their trajectory angle with the axis by a factor called the post-specimen compression. This value is often between 2 and 5 and is easily measured, but is not easily calculated with accuracy; its importance will be seen in Section 4.3.1, where we consider the detectors.

2.4.3 Objective aperture

In the STEM the objective aperture is a beam-defining aperture situated just before the specimen. The STEM objective aperture size directly determines the angle of convergence (see *Figure 4.2*) and the beam current for the probe focused on the sample. Sometimes this aperture is referred to as the 'real' objective aperture to distinguish it from the VOA. Conventional TEM and (S)TEM instruments use the condenser aperture as the beam-defining aperture. Most instruments are based on CTEM technology and have an objective aperture after the specimen in the objective lens back focal plane. The strength of the CTEM objective lens field after the specimen is sufficient to bring the transmitted beams to a focus to form the diffraction pattern (see *Figure 4.11*).

The physical aperture size, in STEM, for obtaining the optimum angle of convergence at the specimen is partly determined by the distance to the sample. Typically the size is around 50 μm diameter and the distance a few millimetres, hence angles are measured in milliradians. Smaller sizes are useful for diffraction contrast in materials with relatively large lattice constant, while larger apertures are useful for lattice imaging. Unlike the TEM, however, high resolution STEM imaging usually does require a real objective aperture.

In the STEM with selected area diffraction mode, the C2 lens relaxes to focus the probe at the objective aperture plane (which is in the objective lens front focal plane), so that the beam is virtually parallel at the specimen. Recall that if an object is placed at the focal point of a converging lens the image is formed at infinity. In SAD the rocking point is at the SAD aperture and hence at the sample too, since the objective lens strength is unaltered when changing to diffraction mode (see *Figure 5.2*). Finally the real objective aperture must be removed from the beam path in order to view the diffraction pattern. If the VOA is used in a STEM it is not visible in the diffraction pattern, because it is situated before the scan coils; it need not therefore be removed.

2.4.4 Specimen holder

Specimen holders vary from manufacturer to manufacturer, most having 'side-entry' arrangements where the sample is held at one end of a long rod (see *Figure 1.1b*). Side-entry systems have great flexibility in engineering design terms since 'services' can easily be fed to the specimen along the rod. Thermal expansion along the rod, however, is a major cause of drift. This thermal drift can be reduced by using 'top-entry' cartridges (which have cylindrical symmetry) which drop into a stage. All the movements of specimen holders inside microscopes are controlled by moving the whole stage or parts of the stage.

The VG603 series STEMs have side-entry stages featuring a 'detachable' rod to reduce thermal drift to the outside. The VG HB5 and 601 series STEMs (with one exception) have top-entry stages, and they combine a high mechanical stability with a low drift rate. These top-entry instruments have, by necessity, a relatively large upper bore diameter for their objective lens to allow the cartridge 'nose' to enter.

In the VG HB5 and 601s the specimen sits within the inner of two concentric tilting gimbals, each operated by a thin tungsten wire. Specimens are held in a recess (just 3.05 mm in diameter and 0.05 mm deep) by a graphite spring or circlip. The gimbals are made from solid beryllium due to its low energy X-ray emission (hence low background contribution). The choice of low atomic mass construction materials (such as beryllium) is fairly universal for AEMs. In the standard VG version there is no height adjustment that would allow the tilted sample to be returned to the same working distance from the lower pole face, though some so-called 'z-shift' stages do exist. Lateral movement of the stage is driven by stepper motors, but is restricted by the upper pole-piece bore diameter and the proximity of the EDXS detector to about ± 1 mm.

The tight restrictions on physical space, in the small pole-piece gap of high performance objective lenses, makes the design of specimen holders critical. Specimen holders need to combine sturdiness with openness, small size with freedom of movement, and 'safety' from accidental touching of nearby surfaces within the lens. The 3 mm diameter of the standard specimen makes tilting through large angles often incompatible with the small gap.

2.5 Post-specimen optics and detectors

Electrons emerging from the specimen and the objective lens magnetic field enter a field-free space often referred to as the 'far-field'. Post-specimen alignment coils and the various electron detectors and spectrometers used in STEM are located in this far-field region.

2.5.1 Secondary electron detector

Low energy secondaries emitted from the exit surface of the specimen in a top-entry type STEM (entrance surface in the (S)TEM) are confined by the post-field of the objective lens (pre-field in the (S)TEM). The magnetic field keeps the secondary electrons in a tight orbital radius until they are released into the low-field region after the lens (before the lens in the (S)TEM) where they are detected. In this low-field region a potential of several hundred volts, applied to a collector, brings the low energy secondaries into a region of much higher electric field (about 10 kV cm^{-1}) where they are accelerated to strike a scintillator. The light generated is a signal that can be amplified and displayed on a monitor exactly as in the SEM.

In most (S)TEMs, the secondaries come from the same surface the X-ray detector 'looks' at (i.e. the entrance surface, see *Figure 3.1*); one disadvantage of the VG601 STEM's geometry is that secondary electron images are usually formed from the opposite side of the sample from which the X-rays are collected. For an instrument to collect secondaries from both surfaces simultaneously it is necessary for the specimen to sit near the maximum of the axial magnetic flux density in the pole-piece gap. Furthermore, special consideration is required to guide the secondaries out through the narrow bores of the objective lens into low-field regions where they might be detected, without unduly affecting the alignment of the mainstream electrons. In the standard VG STEM the electrostatic stigmator in the front bore denies passage to the secondaries (even if they could get over the peak in the magnetic field).

2.5.2 Post-specimen alignment coils

In the simplest STEM systems, that is, those with no further lenses after the objective post-field, there is space above the specimen cartridge for the loading mechanism. The emerging electrons are about to enter the detector region of the microscope and some flexibility in, for instance, steering diffracted beams into the bright field detector is required. The alignment of the objective optical axis to that of the electron energy-loss spectrometer (EELS) is achieved in part by mechanical alignment, but is completed with a double deflection electrical alignment. A bellows-sealed set of coils (often called Grigson coils) is placed on the axis for this purpose and takes up about 10 cm along the beam direction. In principle (but no longer in practice on the VG601s) the emerging scanned beam can be 'de-scanned' back on to the optical axis by these post-specimen coils.

2.5.3 Post-specimen lenses

In all (S)TEMs there are projector lenses that can be used to transfer (and magnify) the image or the diffraction pattern on to a viewing screen, video recording system, or photographic plate. Some STEMs also have post-specimen lenses that provide control over detector collection angles (e.g. keeping camera length constant independent of specimen height) and so maintain a fixed object point for the electron spectrometer. Almost no STEMs exist that routinely allow an image of the sample to be recorded directly on a viewing screen; thus the direct recording of probe size and shape is virtually restricted to (S)TEMs alone.

2.5.4 Annular dark field (ADF) detector

The annular dark field (ADF) detector geometry is similar to that of a backscattered detector of an SEM, but it detects scattered electrons transmitted by the sample. The detector itself is simple enough: an annular scintillator is placed concentrically about the post-specimen optical axis and the light generated by the scattered electrons is collected and amplified using a photomultiplier. The angles of acceptance by the ADF detector depend on the size of the scintillator annulus and on the design of the electron optics, especially in an instrument with post-specimen lenses when the acceptance angles are variable.

Important information about the specimen is obtained from the intensity of the output from this type of detector because it may collect electrons scattered through quite high angles. In the simplest model, electrons which encounter strong scattering centres within the specimen (such as atomic nuclei) are scattered through high angles. There is also evidence that high-angle ADF (HAADF) images (formed using detectors with large inner annular diameter) may have intrinsically higher image resolution than bright field images. This is perhaps easiest to understand by realizing that higher scattering angles result when electrons pass close to an atomic nucleus (exactly like Rutherford scattering) and hence these electrons better pinpoint the location of the strongly scattering nucleus. A corollary to this is the close resemblance of HAADF images to 'background' X-ray maps (see Chapter 7).

2.5.5 Diffraction pattern observation screen (DPOS)

The diffraction pattern observation screen (DPOS) is a thin phosphor (or similar material) screen located at about the same position as the ADF detector. Diffraction patterns formed on this phosphor screen can be viewed, photographed, or recorded with a camera system. The analogue in CTEM is the phosphor screen where virtually all observations are made. (Note that the DPOS is not shown in *Figure 2.1*.)

At low magnifications, in a STEM with no post-specimen lenses, a view of the scanned raster pattern will appear because the scanned beams are not quite parallel to the axis (see *Figure 2.3*). If the electron probe is stationary on a crystalline part of a specimen then a convergent beam electron diffraction (CBED) pattern will be visible. If the scanning system is switched on to scan the probe over this area of the specimen (i.e. high magnification) the CBED pattern will still be visible, with only a slight loss of detail. Thus it is possible (since the DPOS often has a small hole in it) to view a high magnification bright field STEM image simultaneously with the CBED pattern.

2.5.6 Collector aperture

An aperture, placed about the post-specimen field axis, after the ADF detector and the DPOS but before the electron spectrometer and the bright field detector, allows electrons to pass which have undergone low angles of scatter. The collector aperture defines the coherence of the bright field STEM image and is analogous to the condenser aperture in

CTEM when the beam is fully focused on the specimen (see Section 4.4.1). Electron probe lateral (or spatial) coherence is the distance on the specimen across which the phase of the electron wave is essentially constant. FEG systems display high spatial coherence analogous to laser-beam-driven optical systems. Distance on the specimen can be translated into angles in the diffraction pattern via Bragg's law, and we shall see how these are related in Chapters 4 and 5. The collector aperture angle has a central role to play in defining the energy resolution of the electron spectrometer.

2.5.7 *Bright field (BF) detector*

The bright field (BF) detector is an axial detector (scintillator and photomultiplier) usually placed after the ADF detector and detects those transmitted electrons that have undergone low angles of scatter (as defined by the collector aperture). The image acquisition method is essentially serial (each picture point is acquired in turn) and in a digital system, the computer stores the information directly as the signals are generated. The images formed are very similar to CTEM bright field images. Tilted-beam dark field images may also be obtained using post-specimen alignment to 'steer' diffracted beams into the collector aperture. Bright field STEM images can have the same image resolution as bright field CTEM images but are usually very noisy (see Chapter 4, *Figure 4.9c*). In both cases the electron beam current at the specimen may be about the same (say 10^{-10} A), but in STEM only a small portion of the current is collected in bright field while in CTEM most of it is used. If the convergence and collector angles are matched the STEM images will not be noisy but will lack coherence (i.e. BF lattice imaging will **not** be possible).

2.6 X-ray and energy loss spectrometers

All STEMs can (in principle) have an energy dispersive X-ray spectrometer (EDXS) fitted. EDXS detectors are normally placed on the gun side of the specimen in the objective lens, that is, pointed at the entrance surface (as discussed further in Chapter 3, see *Figure 3.1*) and have various capabilities (see Chapter 6 for further details).

Many STEMs are also fitted with an electron energy-loss spectrometer (EELS), located after the ADF, DPOS, and BF detectors. In some systems the BF detector is placed after the EELS system so that energy filtering is possible. An EELS system provides better energy resolution than an EDXS system (to separate closely spaced atomic energy-level transitions) and some advantages in collection efficiency, great sensitivity to structural details (e.g. electronic bonding effects), and no need for any fluorescence correction. Again, see Chapter 6 for more details of this technique.

2.7 Summary

Field emission sources are the preferred electron source in the modern AEM, so that meaningful results may be obtained from ever smaller volumes of material. Electron optical design improvements combined with streamlined analysis facilities have proved effective in solving many challenging materials science problems, while the limits of resolution and information are being established as attainable goals.

The emphasis of STEM instrument design is on probe forming while in (S)TEM it is on image formation. Combining these two features is likely to require an optical system that has a marked symmetry above and below the specimen plane and favours a side-entry stage. The positioning and the functions assigned to the apertures, the modes of operation, the amount of magnetic shielding, and mechanical stability are features that may help distinguish different design philosophies.

Sadly, the sole company making STEMs no longer trades and since they were unique it looks as though these dedicated STEMs will not be a permanent feature of laboratories in the future. None the less, the main TEM companies can now provide almost as good a performance in analytical terms (see Chapter 8 for details). The analytical techniques and the requirements demanded by the specimen itself are universal. In the next chapter we look closely at sample-related issues including specimen preparation, geometry, and beam-induced problems such as contamination and damage.

3 The specimen

A discussion of specimen geometry, which applies to TEM as well as STEM, reveals the main difficulty of transmission techniques: how can such a thin slice of any material really represent the true (bulk) structure? Sometimes the thinnest slices do **not** truly represent the bulk, and analysts are often advised to avoid the thinnest regions, where surface effects are proportionally greatest. This last point emphasizes the need for clean surfaces wherever possible.

3.1 Geometry

In this first section we explore several practical considerations that apply to specimens undergoing microanalysis. We hope to emphasize more of the reasons **why** we do the things we do, rather then just saying **what** we do.

3.1.1 Introduction

STEM specimens are usually just the same as TEM specimens, 3 mm outside diameter and less than 0.25 mm overall thickness. Ideally the specimen used during STEM analysis is about 50 nm thick over the area to be examined, about the same thickness as for high resolution TEM. Reasons for this particular value of thickness are twofold, first, energy loss spectroscopy is best performed on such thin areas of many materials and second, spatial resolution of X-ray microanalysis is spoilt by beam broadening that is comparable to the probe size in specimens about 50 nm thick.

3.1.2 X-ray detectors

Many STEM instruments have X-ray detectors fitted, situated close to the specimen, with relatively large solid angles of detection. *Figure 3.1* shows many of the important features of the geometry of the specimen

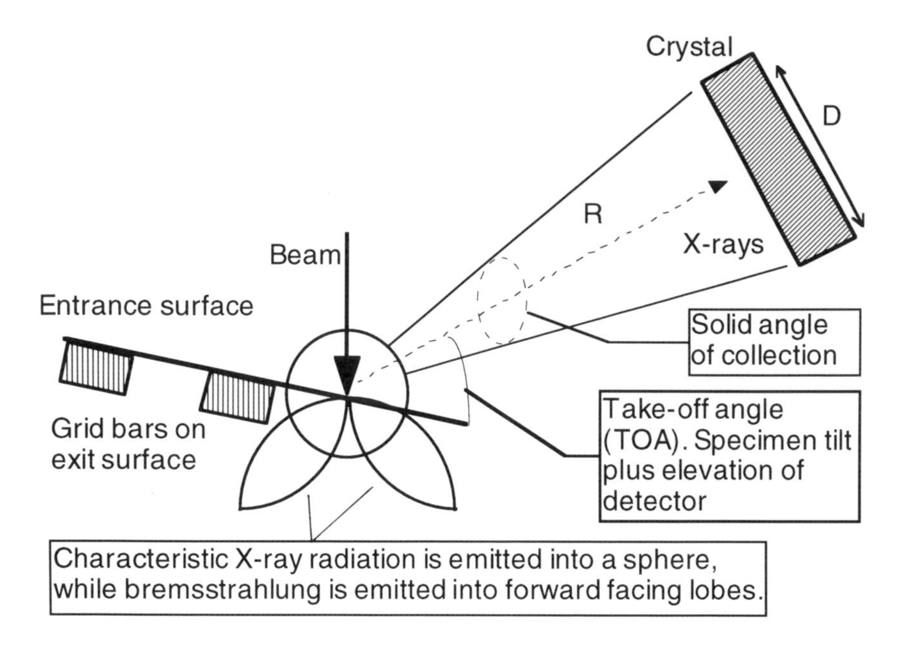

Figure 3.1. Geometry between beam, specimen, and EDXS detector. D is about 6 mm and R is about 12 mm. A sphere contains a solid angle of 4π steradians and has a surface area of $4\pi R^2$. Here the solid angle is $\pi D^2/4R^2$ (0.2 sr). Note the lobes of the bremsstrahlung X-ray background radiation.

in relation to the incident beam and to the X-ray detector. The X-ray detector collects X-rays emitted from, or through, the entrance surface of the specimen. The X-ray detector is located on the entrance side of the specimen because the background radiation (bremsstrahlung) is greatest in the forward direction, while characteristic X-ray generation is uniform in all directions. This forward directionality of the bremsstrahlung background increases with higher accelerating voltages, and therefore instruments operating at higher voltages (200–400 kV) may exhibit higher peak to background ratios.

Specimens supported on grids should be placed in the holder so that the supported film is facing toward the electron gun and hence toward the detector. In this way there is minimal X-ray absorption by the grid bars and the signal is at its strongest. Most specimen holders have a slot cut in the wall surrounding the specimen (see Section 2.4.4), so there is no obstruction to the detector over the entire solid angle of acceptance.

The specimen itself may, as a result of its geometry, cause preferential absorption of lower energy X-rays, so affecting the analysis. The best way to gauge possible geometrical problems that may give rise to X-ray absorption is to carefully consider the spectrum background (see Chapter 6 for details).

When a cross-sectional specimen is made to study interfaces it is usual to orientate the specimen so that the line of the interface is towards the X-ray detector (see *Figure 3.2*). This can be understood by

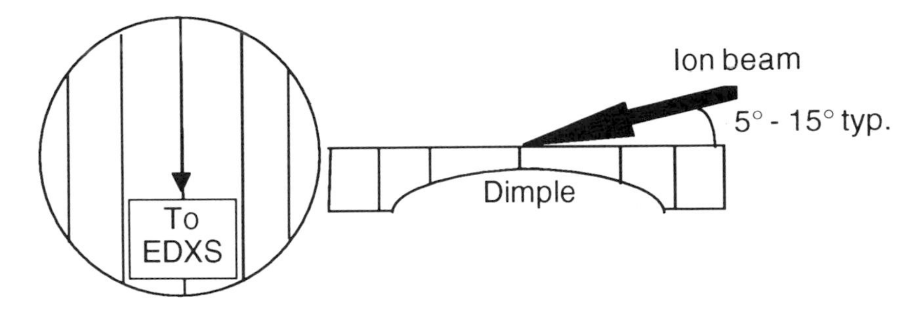

Figure 3.2. Left: looking down on a cross-section of a sample where the material is cut into strips that are glued together (note orientation for EDXS), and right: dimpled on one side and ion-milled.

thinking about the absorption correction that may be applied to the analyses from the different layers in the specimen. If we have silicon on top of germanium or iron on top of aluminium and look through the cross-section we will see the layers side by side. When analysing the layers it is best to keep the composition of the material through which the X-rays to be detected must travel the same as the material that is being analysed. Otherwise we might generate X-rays in silicon and absorb them in germanium as in one of the examples above. The normal analysis program (which applies corrections for absorption and fluorescence) would be unaware that the germanium was even there and would apply absorption corrections relevant to silicon.

The moral of this scenario is to keep sources of error or uncertainty to a minimum whenever possible. In general choose a point of analysis that minimizes absorption and fluorescence effects; find out where the detector is in relation to the image, and pick parts of the specimen that have least material between that part and the detector.

3.1.3 Specimen preparation

An example of a cross-sectioned specimen is shown in *Figure 3.2* and the following discussion addresses several practical points in specimen preparation. The interface in the centre is the one of interest, while those away from the centre are sacrificial and may be chosen from other materials. Better results are likely to be obtained if a low angle ion-mill can be used in conjunction with a dimple that leaves only about 20 μm in the centre. Ideally, specimen thickness should be less than 100 μm overall, otherwise the low angle ion-mill cannot polish the centre of the dimpled side, to remove the mechanical polishing damage. We should load this specimen with the dimple away from the detector otherwise it may be necessary to tilt towards the detector to remove shadowing effects. This tilt may be necessary if the detector geometry accepts X-rays travelling in the plane of the untilted specimen (i.e. at 90° to the beam axis) as can happen when the solid angle is large and the take-off

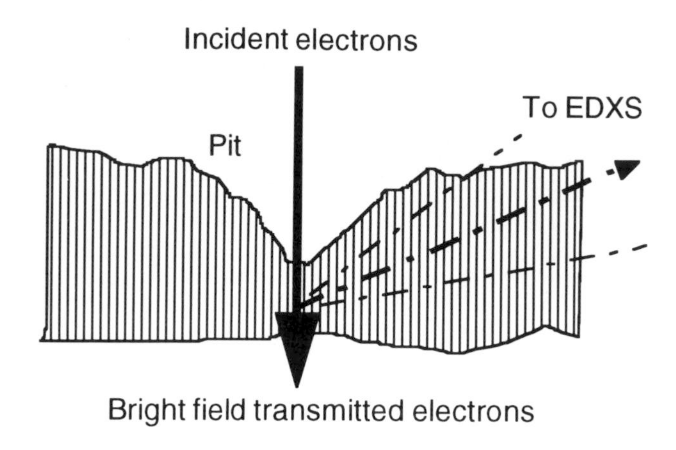

Figure 3.3. An unfortunate (but typical) etch pit geometry.

angle is low such that the cone of acceptance (see *Figure 3.1*) grazes the specimen.

The microscopist should be aware of problems that might arise if the analytical signal of interest (be it X-ray or secondary electron emission) has a very different geometrical path to follow before detection from that of the main transmitted electrons. Consider as an example, shown in *Figure 3.3*, an etch pit that is very deep relative to its width, perhaps due to preferential etching of one phase relative to another or as sometimes seen with electropolished metallurgical specimens. The transmitted beams might give the impression of thinness at the point of analysis (at the bottom of the pit), indeed a thickness measurement is likely to show a relatively small value here compared to the regions surrounding the pit. Often these pits fill up with unwanted material during polishing or coating processes.

In this case the EDXS signal is always going to be absorbed by a large thickness of material compared to the thickness of the irradiated volume (unless the specimen can be turned over). Detector geometry is vitally important if a quantitative statement is to be made about composition at the point of analysis in a specimen where the geometry is not simply parallel sided, semi-infinite, and of uniform composition (see *Figure 6.4*).

3.2 Tilting possibilities

It is nice to have large tilt angles available when studying crystalline or three-dimensional objects, but in a microscope there are limitations placed on the range of angles by the amount of space between the specimen holder and the lens. Specimen tilt is useful for analysis because

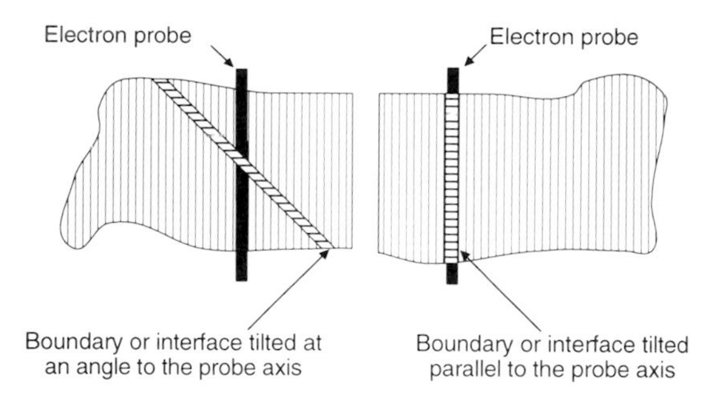

Figure 3.4. Two tilt orientations. When the boundary is parallel to the beam a greater sensitivity of detection of the segregant is obtained.

maximum detection sensitivity is obtained when the incident probe passes along the length of the object under study. For example, a grain boundary may have a small concentration of an element segregated to the boundary, and if the boundary plane cuts across the line of the probe, only a few atoms will lie within the probe (see *Figure 3.4*). Tilting the specimen such that the grain boundary lies parallel to the probe direction will substantially increase the number of atoms lying inside the probe (especially if the probe is a small one). Indeed the result can be startling and well worth the extra effort needed to achieve an accurate alignment. Of course it is also valid to tilt the beam along the boundary, but this is not the usual way as it can cause problems with alignment and image quality.

Figure 3.4 shows two boundaries (one inclined, the other normal) within a specimen that is perhaps 50 nm thick. If the (ideal) electron probe is 2 nm diameter (cross-sectional area 3 nm^2) and the boundary is only two atom layers thick, the number of atoms probed at the inclined boundary (see *Figure 3.5*) is likely only to be about 50 (if the atomic spacing were 0.4 nm). The number of atoms inside the probe when the boundary is parallel to the probe is about 30 times as many (about 1500 atoms of interface material) and is thus more easily measured.

Tilting the specimen is likely to change the height of the region of interest inside the objective lens and may lead to a defocused image. Although it is easy to refocus the lens, many instruments offer an alternative, to change the height of the specimen using a '*z*-shift'. The latter method is to be preferred because the lens focal length remains unchanged, so the magnification and diffraction calibrations and probe properties are kept constant. Other disadvantages associated with changing the lens strength instead of the height of the specimen (or its holder) are that

- changes in the optical transfer to subsequent devices such as the energy loss spectrometer result in loss of EELS energy resolution;
- change in the heat input to the lens body through the lens windings leads to thermal drift;

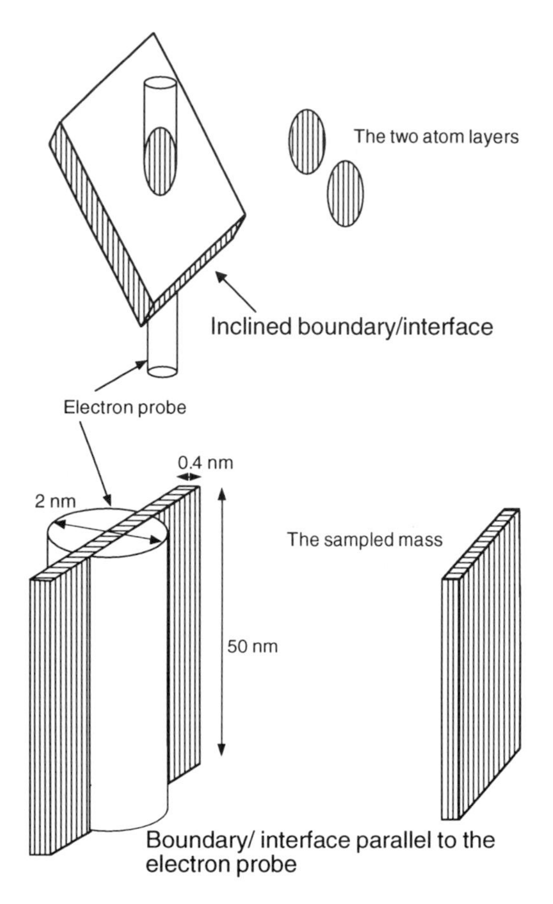

Figure 3.5. A simple model to estimate analysed volume. With the electron probe at an angle to the boundary or interface of interest (upper part), the 2 nm diameter (idealized) probe touches about 25 atoms as it enters the interface material (if the interatomic spacing is about 0.4 nm). Since there are just two atom layers at the boundary/interface the total number of atoms encountered is about 50. With the probe and boundary or interface parallel, however, if the interatomic distance is 0.4 nm, then the idealized probe touches about 12 atoms in each layer. There are about 125 atom layers for a specimen 50 nm thick, so the total number of atoms 'sampled' by the probe is about 1500.

- unwanted geometrical effects may occur with respect to the collection of X-rays (e.g. the analysed volume may move into shadow) since the plane of the specimen, probe axis, and X-ray detector axis no longer coincide.

If allowance is made in the microscope lens design to provide large scale tilting and shifting capabilities as well as z-shifting adjustments, then the lens gap (the distance between magnetic poles) is necessarily increased. Large-gap lenses have weaker (and so poorer) focusing properties than small-gap designs, and since lens aberrations increase with focal length, the ultimate resolution is reduced. A compromise is always found, usually by restricting the specimen tilt.

3.3 Airlock

A good vacuum in the microscope is a high priority in STEM for several reasons: electron emission current stability (especially for cold FEG systems), low rate of contamination, and windowless EDXS compatibility. In order to preserve the quality of the vacuum in the microscope over long periods and during many specimen exchanges, an efficient airlock mechanism is required. Some modern AEMs have a bake-out facility to improve the specimen chamber (or column) vacuum, but only in VG STEMs can the entire instrument be baked at temperatures over 130°C. System bake-outs take place as required, normally overnight and several hours must elapse before restarting after a bake. However, even a super-clean microscope is not immune to problems that can be introduced with the specimen itself, and some kind of specimen pretreatment is often employed.

The type of pumping system used with most (S)TEMs is an ion getter pump often combined with some cryogenic shielding surrounding the specimen. The object is to keep the vacuum as clean as possible, with respect to hydrocarbon vapours, and perhaps more importantly to keep the internal surfaces clean. Techniques that help maintain optimum conditions for STEM are scrupulous care whilst handling specimens and holders (wear clean gloves) and with specimen preparation techniques and storage methods (e.g. do not use gelatine capsules).

It is preferable to use clean and dry specimens, though sometimes specimens are stored in 100% ethanol (not acetone) to prevent oxidation. Take care to store specimen holders in a clean, dry place (a vacuum desiccator is probably best) since surfaces accumulate moisture that slows down the pumping system. If dirty specimens do get into the microscope then the time between bake-outs will be shortened, and this 'down-time' becomes inconvenient to other users. Heating specimens with a lamp can speed up desorption of vapours, though the heat will take time to dissipate and leads to specimen drift. Care is also required because some specimens cannot be heated with impunity, they might melt or undergo an irreversible change. Some adhesives used to glue parts of the specimen together or attach a support ring can exhibit high vapour pressures that might cause contamination upon heating in the microscope.

The specimen holder is sealed in the airlock and evacuated to around 10^{-2}–10^{-5} mbar before introduction into the analysis chamber (column vacuum) of the microscope. Side-entry holders have an 'O-ring' vacuum seal and special care must be taken with this seal to ensure consistently good performance. The airlock need not have a large volume, and so need not have a large pumping system attached. If a simple oil-filled rotary pump is used, avoid pumping the airlock for more than a few minutes, otherwise oil vapour back-streams into the airlock. Ideally a diffusion pump or turbomolecular pump should be used and then the evacuation can proceed to high vacuum of $< 10^{-5}$ mbar.

3.4 Contamination

Contamination is a problem in nearly all analytical electron beam instruments but can be avoided by careful specimen preparation, storage, and handling. Winning the battle against unwanted carbon contamination is perhaps the greatest challenge to instrument manufacturers and regular microscope users. The problem often arises when mobile hydrocarbons diffuse across the specimen surface to the point of analysis, where the electron beam cracks them into their constituent atoms. The carbon deposits as amorphous contamination spots above and below the specimen, building up faster than any electron-beam sputtering of carbon from the surface. The chemical composition of the contamination is about 90% carbon plus about 10% oxygen (if hydrogen is present it cannot be detected by EDXS or EELS). Evaporated carbon films seem to have a similar composition.

Contamination can cause beam spreading even before the probe has entered the region of real interest, thus enlarging the analysed volume and hence degrading the spatial resolution of the analysis. X-rays may be absorbed by the carbon in the contamination spot, so reducing the accuracy of the chemical analysis of light (low atomic number) elements. Analysis of carbon itself in a material when contamination is present creates obvious difficulties. Removing the background signal arising from a build-up of carbon is difficult, and often produces an inaccurate result. The microscopist's time is better spent eliminating contamination at source, before the specimen is loaded in to the microscope.

One method is to fix the contamination in place so that it cannot migrate to the area under investigation. The specimen is flooded over a large area, hundreds of μm square, with electrons or powerful UV light. The object is to crack the mobile hydrocarbons on the surfaces of the specimen and cause them to become immobile or even absent (desorbed). When this works it can last for hours, but if it does not work then the microscope vacuum itself may be to blame. A poor vacuum system contains significant numbers of hydrocarbon molecules and this adds to the contamination build-up. Such a system needs a good bake-out and if problems persist it is necessary to clean the specimen holders. A vacuum leak is a possible source of contamination, though on occasions a water vapour leak actually removes the carbon contamination as rapidly as it forms – it can also strip away carbon from the specimen itself!

Contamination is worst in systems that have a high beam current and small probe size, as is found when converting from tungsten to lanthanum hexaboride or FEG. The higher current density of the smaller probe rapidly forms contamination spots, which may be useful as markers or even for estimating specimen thickness, but preference is for no contamination. Field emission gun systems are the most susceptible to contamination, but with best practice during manufacture and in service there need not be contamination with clean specimens. Clean apertures are important too, especially if they are beam defining since desorbed hydrocarbon molecules can travel along the beam direction to the specimen.

3.5 Beam damage

Beam damage comes in two main forms, direct displacement of atoms by electrons (momentum transfer) and ionization damage (bond breaking). The former is most important above a certain threshold accelerating voltage (e.g. 150 kV for aluminium) while the latter is most often a problem with oxides. Ionization damage can release oxygen atoms from their bonds and if they are mobile within the structure of the specimen, they can collect together as bubbles. These bubbles grow visibly under the beam and eventually burst out through the surface leaving a hole in the specimen. The result is often called hole drilling and has some potential as a storage medium (like a laser cutting pits in a CD). The rapidity of the damage mechanism is a function of the electron density within the beam and since the FEG system has the highest electron density it causes the most damage this way. See Section 8.6 for further examples.

Beam damage is always going to be a problem and there are very few things that can help reduce its effects without changing the very structure and/or chemistry we are trying to study. One thing that can sometimes help with oxides is to coat one of the surfaces with carbon. This reduces the charging phenomenon that is a nuisance anyway, also, the oxygen ions are more likely to recombine if the charge field is neutral. There is evidence also that a carbon coating on the electron exit surface can reduce the rate of atom sputtering from the surface by momentum transfer. FEG-STEMs with particularly clean vacuum systems are in fact more susceptible to this sputtering process than systems whose vacuum actually deposits carbon on the specimen's surface.

Finally, the higher voltage machines (200–400 kV) are prone to all types of beam damage but ionization damage ought, in principle, be less important for a given number of electrons, since the ionization probability decreases with increasing accelerating voltage. Weighing against this trend, however, is the increased displacement damage (higher momentum to transfer) and the fact that higher voltage electron sources are brighter (i.e. have more electron current density per unit solid angle in their probe).

In the next chapter we consider some examples of the STEM's imaging capabilities, examine some of the relationships between STEM and TEM, and explore a few of the limits.

4 Imaging in the STEM

4.1 Introduction

We easily recognize scanning electron microscope (SEM) images, because SEM images have a depth of field and a three-dimensional quality that makes them look like solid objects. We tend to recognize solid objects by their outline, as this is most clearly related to our sense of touch. The way that light is reflected or transmitted by a solid object helps us decide the nature of an object. The same processes of recognition occur with microscopy, both light and electron optical microscopes. Provided the object is thin enough for transmission we may obtain information from within the object.

By the *image*, we mean the *information* conveyed and stored photographically or electronically. With transmission techniques it is hoped that this information might have little to do with surface topography or shape. Confidence in the interpretation of an image from an unknown structure is established by reference to known structures and their images. It is the experience of the microscopist that translates these images into the possible three-dimensional projections of objects. Since the specimen is thin and transparent, density and crystal orientation variations within the specimen have to be considered when interpreting the image.

Like an SEM, a STEM forms a focused electron probe (a demagnified image of the electron source) using the C1, C2, and objective electron lenses. The probe is defined by various apertures and moves over a small area, in a raster pattern, on the specimen and the electronic signals received from detectors **beyond** the thin specimen are amplified. The signals are used to modulate the brightness of cathode ray tubes scanned in exact synchronization with the scan on the specimen to form images of the specimen. The detectors may be sensitive to light, electrons, X-rays, current, or any detectable signal. The signals may also be stored in a computer frame store which can be updated before display as an image on a monitor or saved to a file on a computer disk.

In STEM we measure simultaneously, point by point, **signals**, each containing different information about the specimen. Examples are the

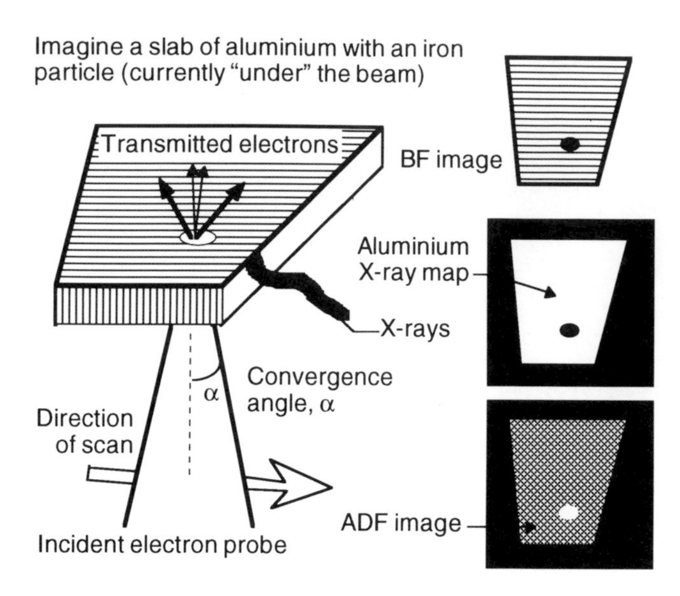

Figure 4.1. A schematic diagram showing how points on the object correspond to points in the images seen on the monitor screen.

BF signal, the ADF signal, the Fe X-ray signal or the Ca EELS signal; and we might display, pixel by pixel, each of these signals as **images**. Often the diversity of information from these multiple signals can help or simplify our understanding of a specimen, ideally making for a more complete description.

Figure 4.1 shows a schematic illustration of the method of STEM image formation; think of the different images as being in different channels.

4.2 General discussion of probe forming

Images formed with transmitted electrons contain information about specimen structure and composition. However, in the STEM especially, at low magnification, there may be some contributions from the detectors themselves forming a kind of background pattern that needs to be understood **before** it can be ignored. An image is a convolution of specimen and instrument functions; the position dependence of the image alone is what helps us separate these two factors, hence moving the specimen helps you perceive its contribution to the image.

The size and shape of the probe in a STEM are determined by the source, beam-defining apertures, the electron energy, and the settings and quality of the probe-forming lenses. The influence of these factors on the image is complicated by the wave-like nature of electrons, and by the way the above factors interact. The probe's shape and position are very susceptible to interference both from mechanical vibrations (including

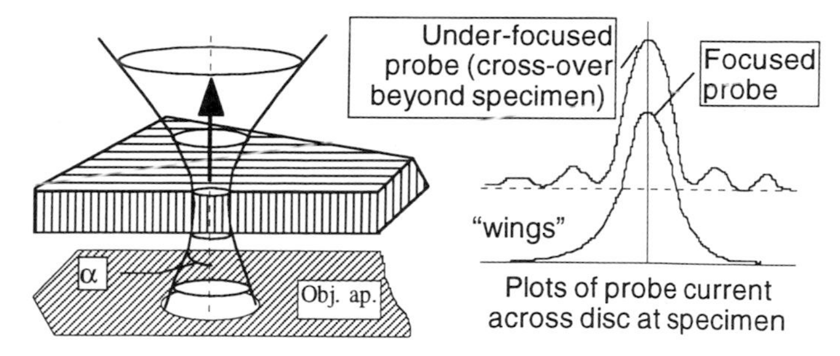

Figure 4.2. The probe converges with a semi-angle, α, and is focused to reach a minimum diameter at the specimen itself. Profiles of current density illustrate one effect of focus on the probe shape. The 'wings' are due mostly to aperture diffraction and lens effects such as spherical aberration.

acoustic) and from stray electric/magnetic fields, resulting in the images of sharp edges being ragged.

Surprisingly, probe size and probe current are not independent of each other. Given an original number of electrons emitted from the gun you might think there should be the same number inside all probes. Unfortunately the method used to make small probes does not conserve the original number of electrons, and a huge proportion is lost on apertures. The focused probe shape (envelope) is mostly defined by the beam-defining aperture angle, α, while the density of electrons within each disc in *Figure 4.2* has approximately the same shape as a normal distribution curve (a Gaussian function). The current density inside the probe at the specimen has a ringed (Airy disc) structure when the probe is focused beyond the specimen plane (an 'under-focused' probe would then strike the specimen). Images formed with different probe shapes have different contrast. An ideal, but unattainable, probe would have all the current within a certain (small) diameter.

4.2.1 A few numbers and formulae (facts and figures)

The following list of bullet points is not intended for readers who are new to the field; rather it is those who have some experience in estimating instrument resolution who are most likely to follow the numbers through. However, a novice might still find it worthwhile, since everything is self-consistently defined. *Figure 4.3* shows a schematic ray diagram with two settings of the condenser lenses in a STEM with a virtual objective aperture (VOA) in use.

- Electron wavelength, λ, after 100 kV acceleration is 3.7 pm (0.037 Å).
- Typical STEM probe convergence semi-angle, α, at the specimen is 10 mr (0.01 radians).
- Spherical aberration limit on probe size formed from a point source is about $0.25 C_s \alpha^3$ (e.g. 0.25 nm if objective lens spherical aberration coefficient, $C_s = 1$ mm).

- Diffraction limit on probe size is about $0.61\ \lambda/\alpha$ (e.g. 0.23 nm at 100 kV).
- Gun brightness for a 100 kV cold FEG is about 10^9 A cm^{-2} sr^{-1}.
- Electron beam current density, J A cm^{-2}, is gun brightness times beam solid angle.
- Solid angle, Ω, is $2\pi(1-\cos\alpha)$; for small angles it is $\pi\alpha^2$ (e.g. 0.3 milli-steradians).
- Current density in a FEG-STEM probe is about 300 000 A cm^{-2}. This far exceeds the current density limit for copper wire (about 155 A cm^{-2}).
- Analytical beam conditions (i.e. a weak C1 lens) arise with source demagnification of about 10. The angular magnification is also 10, so $\alpha_2 = 10\alpha^\dagger$ (see *Figure 4.3*).
- FEG source size is about 5 nm and angle of analytical beam leaving tip is about 1 milli-radian (defined by the VOA).
- Geometrical source size contribution to analytical probe is about 0.5 nm.
- Gun chromatic aberration coefficient, C_c, is about 10 cm and extra-high tension (EHT) stability (including source energy spread for a FEG) is 5 p.p.m. (0.5 V in 100 000 V).
- Gun chromatic aberration is $C_c\alpha^\dagger\Delta E/E$ and contributes about 0.5 nm to analytical probe size ($\Delta E/E$ is the stability of the EHT and $\alpha^\dagger$ is the angle at the tip).
- Lower C_c for the gun and a sharp tip (small source) allow a smaller analytical probe size.
- Ultimate resolution is obtained with a source demagnification of about 50.
- Lower C_s for the objective allows a larger convergence angle to be used for a given probe size, hence more current and better ultimate resolution.
- Gun brightness for a given accelerating potential is constant and the probe angle determines the beam current density.

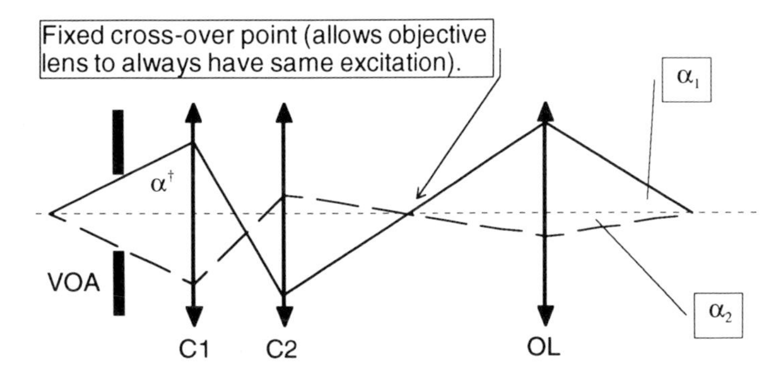

Figure 4.3. A simple ray diagram illustrating demagnification. Case 1 (solid line) has a strong first condenser, a large angle of convergence, and a small source size contribution to the probe. Case 2 (broken line) has a weak C1 lens. The illumination arriving at C2 has higher current density (covers a smaller area). The angle at the sample is smaller than in case 1. The source is less demagnified since the C2 lens is more magnifying than in case 1. Note: the diagram is not to scale.

Summary of typical values:

- spherical aberration limit ≈ diffraction limit ≈ 0.25 nm;
- analytical probe conditions: source size ≈ gun chromatic aberration limit ≈ 0.5 nm;
- high resolution method for STEM: demagnify the source and gun aberrations, leaving diffraction and spherical aberration limits.

4.3 Processes in image formation

There need not be an image **within** the STEM, usually it is only on the monitor. This monitor image is a two-dimensional display of picture elements (pixels) with typically 256 (8-bit) displayed grey levels of intensity, although often more bits are stored. In STEM the image is collected serially, that is, pixel by pixel, whereas in the conventional TEM it is collected in parallel, that is, all pixels at the same time. The interesting part, so far as the microscopist is concerned, comes during the interpretation of the image in terms of the supposed structure and composition of the object.

The individual pixel intensities (in a digital system) are found by a chain of rapid processes starting with a typical STEM detector:

- high energy electrons are converted into light by striking a scintillating material (e.g. yttrium–aluminium–garnet or YAG);
- the light is guided out of the microscope as efficiently as possible;
- the light is converted back into electrons at the photocathode of a photomultiplier tube (PMT) and the number of these electrons is greatly multiplied by the action of the PMT;
- finally, after analogue amplification, the electronic signal is digitized and held in frame store memory.

The dwell time of the beam at a particular point on the specimen, derived from the rate at which the information is displayed or updated, can be as short as a fraction of a microsecond or as long as a millisecond. At 'TV' rate there are usually 25 or 30 interlaced frames, each of about 250 000 pixels, acquired and displayed each second! The PMT is indiscriminant – when a flash of light occurs on the scintillator the system simply integrates the total intensity of the signal received.

4.3.1 STEM detectors

There can be as many different STEM images as there are different detectors and detector geometries. The most common detectors, and hence images, are shown in *Figure 4.4* and listed below (collection angles refer to the specimen plane after accounting for lens effects, e.g. electrons passing through additional post-specimen lenses before being detected).

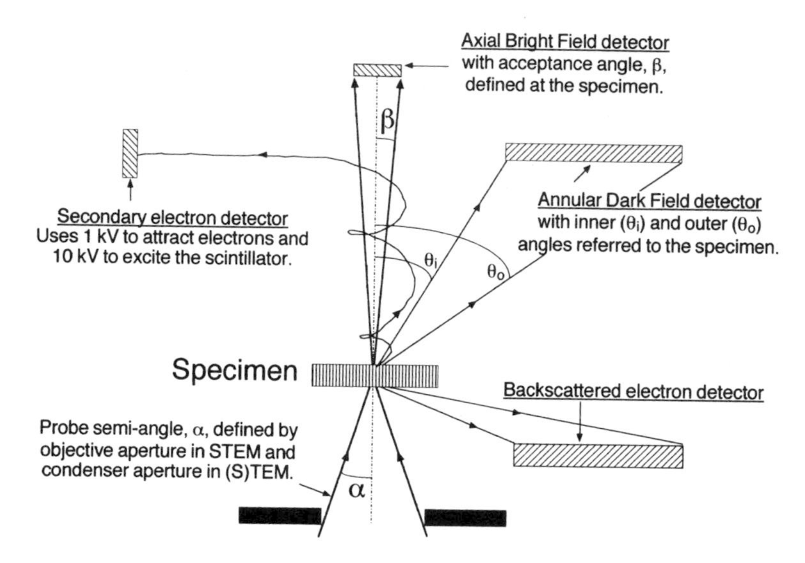

Figure 4.4. Some typical STEM detector geometries (not to scale).

- **Bright field** (BF): axial detector, typically $\beta \approx$1–5 mrad, defined by the collector aperture size and any post-specimen lens effects.
- **Annular dark field** (ADF): inner hole usually just misses collecting the outer edges of the probe envelope, often θ_i–$\theta_o \approx$ 15–150 mrad. Masks are available to alter θ_i in the absence of post-specimen lenses.
- **High angle annular dark field** (HAADF): the large inner hole diameter allows most diffracted beams through, typically θ_i–$\theta_o \approx$ 50–150 mrad.
- **Secondary electron** (SE): low energy electrons without line of sight to the specimen, angles not normally quoted. In principle large angular ranges are possible if care is taken to guide the electrons out through the bore of the objective.
- **Backscattered** (BS): annular detector on the entrance side, θ_i–$\theta_o \approx$ 400–1200 mrad (assuming detector has i.d. 5 mm, o.d. 25 mm and is 5 mm away).
- **Diffraction** (DPOS): shows the angular distribution of transmitted electrons. θ_i–$\theta_o \approx$ 0–120 mrad. Sometimes there is a small hole in the DPOS with $\theta_i \approx$ 0.5 mrad.

Other images may yield a variety of chemical maps. Detection of emitted characteristic X-rays gives 'bulk' chemical maps (see Chapter 7) while detection of Auger electrons gives near-surface chemical maps. In addition, detection of visible light gives cathodoluminescent images; and detection of absorbed current measures electron-beam induced conductivity (EBIC). The list is growing with new detector geometries being tried in the search for more information about materials.

There are specially configured detector geometries that yield specific information, or enhanced resolution and contrast, such as, for example, edge detection with a thin annular detector (inner cut-off angle, call it β_i, smaller than α). It is useful to have control over the divergence angle of the beam that arrives at the detector, so that a particular contrast effect can be optimized. Magnetic domain images can be formed with a quadrant type annular dark field detector; the signal from each quadrant can be added to or subtracted from the others to yield information about magnetic deflections within the specimen. The ultimate detector would probably be an array of small detectors that could be individually selected (or masked) to provide almost infinite flexibility in the composition of the image. Unfortunately such a detector is still a little way off, though current work on microchannel plate technology should have a big impact.

4.4 Some typical images from a STEM

Some typical images are shown in *Figures 4.5–4.8* and explained in the captions.

4.4.1 Bright field STEM images

Bright field STEM images are the closest to conventional TEM images, and it is possible to record all the usual types of images (dark field, weak beam, phase contrast, lattice image, Fresnel contrast, and Lorentz contrast) as in the TEM. The collector aperture size in STEM defines the collection angle β (see *Figure 4.4*), and it is this angle that controls the degree to which bright field STEM images resemble TEM images. In the TEM the equivalent angle is defined by the angle of convergence, α. Thus images taken with small STEM collector apertures (e.g. *Figure 4.9c*) resemble TEM images taken with small beam convergence, and in both cases the imaging conditions are said to be 'coherent'.

4.4.2 Annular dark field STEM images

In addition to BF there are the characteristic STEM annular dark field (ADF) images that arise from angular integration of the scattered intensity over a range of angles (from inner cut-off to outer cut-off). Equivalent images are obtainable in TEM using a hollow cone illumination technique either with a special condenser aperture (known as a strioscopic aperture) or a rotating scan of tilted illumination. Examples of STEM images are shown in *Figure 4.9* using different detectors with different angles as indicated. The material is a catalyst-covered magnesium oxide cube.

A low angle ADF image such as is shown in *Figure 4.9b* (inner cut-off only marginally greater than the beam convergence angle, α = 11 mrad)

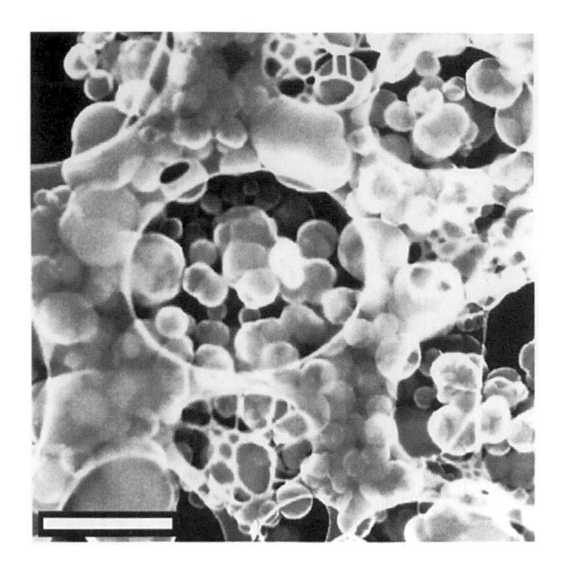

Figure 4.5. Secondary electron image of carbon soot particles supported on a holey carbon film. The image is formed using electrons collected from the exit surface of the specimen and shows that many of the particles are located on the entrance surface of the carbon support film (which in the VG STEM is below the film). We 'look down' on to the sample and the electrons are coming up towards us. Bar = 1 μm. (Courtesy of Dr A. Burden, University of Surrey.)

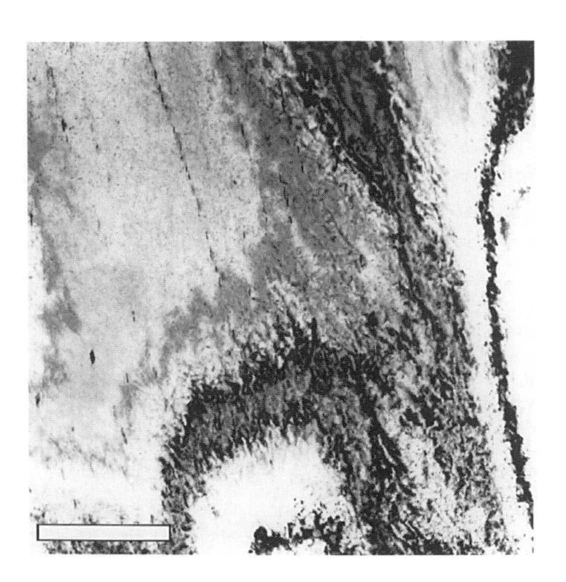

Figure 4.6. Bright field STEM image of a ferritic steel after creep testing at 580°C for 700 h under a stress of 180 MPa. There are many needle-like vanadium carbides aligned within the ferrite grains that may be responsible for the observed embrittlement. Notice the normal bend contour diffraction contrast you can get with a small collector aperture. Bar = 0.2 μm. (Courtesy of Ms S. Finke, University of Manchester.)

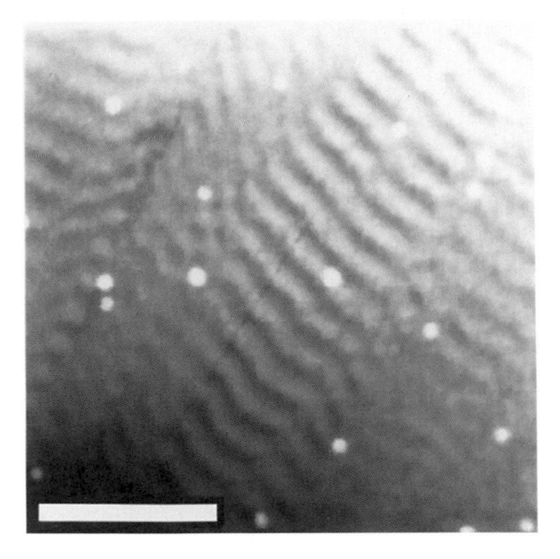

Figure 4.7. Annular dark field STEM image of a cross-section of diamond-like carbon film grown on a titanium carbide substrate. The wavy layers are variations in titanium concentration and the bright features are particles of copper deposited during specimen preparation. The image exhibits *Z*-contrast and the track left by the electron beam during an X-ray linescan (see Chapter 7) is visible as a thin dark line (some mass loss occurs during the analysis). Bar = 0.1 μm. (Courtesy of Dr R. Touaitia, University of Northumbria at Newcastle.)

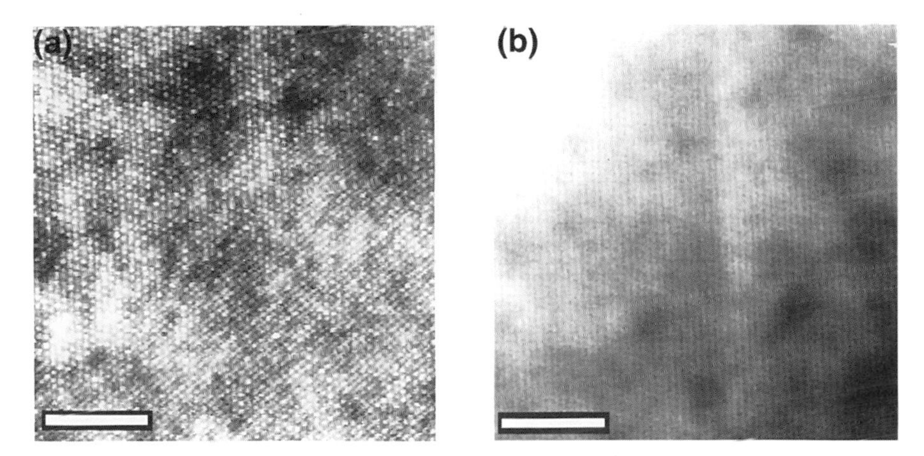

Figure 4.8. (a) Bright field lattice image of a rough silicon/silicon–germanium interface, seen here in cross-section. The interface is hardly visible. (b) High-angle annular dark field STEM lattice resolution image. The lighter region is the higher atomic number Si–Ge. The interface is atomically abrupt according to the sudden change in contrast across the layer. Bar = 4 nm (0.004 μm). (Courtesy of Dr U. Bangert, UMIST.)

shows diffraction contrast quite clearly because strongly excited diffracted beams are being detected. High-angle ADF images, such as in *Figure 4.9d*, show little or no diffraction contrast since the inner cut-off angle of the detector is so large that diffracted beams (largely damped out by lattice vibrations) do not contribute, leaving only thermal diffuse

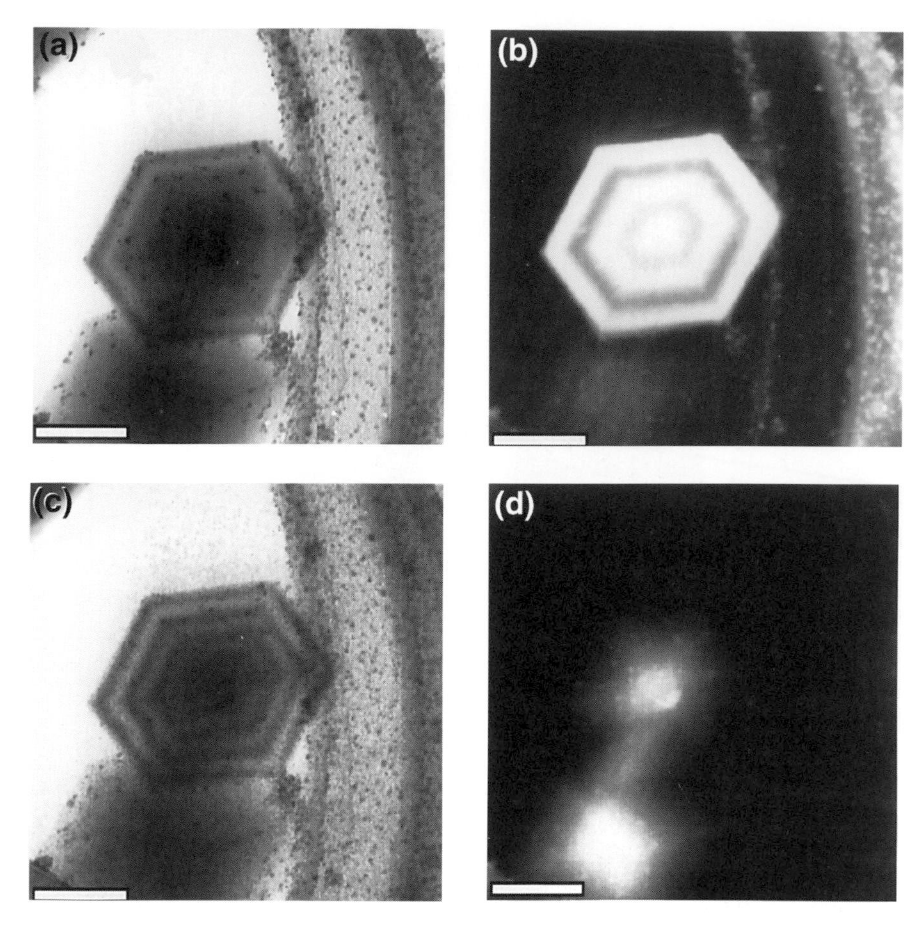

Figure 4.9. A set of images of MgO taken with different STEM detectors: (a) bright field (β = 1.7 mrad); (b) annular dark field (θ_i = 14 mrad); (c) bright field (β = 0.3 mrad); and (d) annular dark field ((θ_i = 47 mrad). Bar = 100 nm.

scattering to be recorded in the HAADF image. There is evidence that the HAADF image exhibits an 'incoherent' nature, since it has none of the thickness fringe effects (see *Figure 4.9a*, *b* and *c*).

4.4.3 Secondary electron and Auger images

An example of a secondary electron STEM image is shown in *Figure 4.5*. Secondary electrons are generated throughout the irradiated parts of the specimen by electron–electron interactions. The low kinetic energy of the secondary electrons means that only those near the surface may escape to be collected. Very few instruments have the capability to collect secondaries from both surfaces of the specimen.

Auger electrons are generated when inner-shell excited ions relax (see Chapter 6). Auger electron emission and characteristic X-ray emission are alternative relaxation mechanisms and both will occur within any electron microscope. Auger electron images and spectra are chemically sensitive because the energies of the electrons (typically between 100 and 2000 eV) are characteristic of the energy levels involved in the transition to the ground state.

4.5 Resolution

In this section we shall discuss image resolution, that is, the level of detail revealed by the microscope, measured in nm. Other definitions of resolution exist in diffraction (Chapter 5) and microanalysis (Chapter 6).

4.5.1 Introduction

A good question is 'what limits the microscope's resolution?' This question is perhaps best answered by experiment; although there is plenty of theory too. Normally we quote a figure which indicates the separation of two points in the image which are clearly distinguishable. This is expressed quantitatively by the Rayleigh criterion [see, for example, Hecht (1990); Jenkins and White (1981); Longhurst (1973), or any other optics textbook]. The limit turns out to be a fundamental one, which derives from diffraction at an aperture, and as a result smaller apertures produce larger images from point objects. Wavelength of radiation (in this case electron wavelength determined by the electron energy) is involved in the so-called diffraction limit to resolution; and shorter wavelengths (corresponding to higher energy electrons) are less affected by diffraction than longer wavelengths.

4.5.2 Defining resolution

If we knew beforehand what the individual features should look like (e.g. how big they are and how sharp are their edges) this information would help us resolve the points and their positions better. A useful concept in this situation is the *point spread function* of the instrument, or the resulting image of an infinitesimally small yet powerfully scattering object. As far as electrons are concerned such an object could be an atomic nucleus. A working definition of ultimate STEM image resolution might be the probe size, the point spread function, or an image width measured from a single atom.

In practice the STEM resolution is the interpretable detail in an image (or a chemical map), which is often the probe size modified by scattering in the specimen. Let us first consider what influences the initial probe size, that is, the diameter within which a fixed proportion of the total beam current lies. We ought to choose this approach rather than the

FWHM, since it has somewhat more meaning for microanalysis. Many of the factors that influence one definition also affect the other, but we ought to be aware of the difference when choosing the size of the probe-forming aperture to use for a given experiment. In microanalysis we often want more beam current than is possible to fit inside the smallest probes; and we accept the inevitable degradation in probe size because we require a rapid and statistically significant analysis of our region of interest. The method of getting more beam current is to reduce the demagnification of the system and so produce a geometrical probe size that is larger than the ultimate one.

4.5.3 Limits to resolution

A word about ultimate STEM resolution: it is the combination of diffraction-limited resolution and spherical-aberration limited resolution. The other limits can, in principle, be eliminated by demagnification, adequate shielding from interference, stability in power supplies, and thinness of specimen. Ultimate STEM resolution is what you get when you reduce the source size by demagnification (this in turn makes gun instabilities irrelevant) and eliminate chromatic aberration effects from the objective lens. If we plot a graph that shows the way different effects of aberrations and diffraction change with aperture angle (in STEM we mean convergence angle) we find that resolution is diffraction limited below a certain optimum aperture angle and spherical aberration limited above this value (see *Figure 4.10*). Source size does not usually figure in these considerations but **we** must consider it because real world microanalysis demands a different choice of aperture for optimum results.

Spherical aberration varies with the cube of the convergence angle, while diffraction is inversely proportional to the angle, as *Figure 4.10* shows. Chromatic aberration is directly proportional to convergence angle, while geometrical source size is dependent on demagnification and so reduces as the angle increases. We illustrate two situations, one where a large beam current (and hence large source size) is the dominant factor (labelled 'A' in *Figure 4.10*); and the ultimate resolution case (labelled 'B'). In the first case we see that the probe is source-size and spherical-aberration limited, while in the second case the source has been demagnified out of the picture and the effects of diffraction have become 'exposed'. We can not operate below the diffraction curve. Different VOA sizes give rise to the different geometrical source size curves in *Figure 4.10*, larger VOAs allow electrons to enter the column from larger regions on the tip. The form of the source size curves is found from a relationship between brightness, source size, and solid angle (namely, source size is inversely proportional to α). The VOA size that corresponds to the diffraction curve ought to be the optimum choice.

Unfortunately a spherically aberrated probe has a substantial disc of minimum confusion and the cross-sectional view of the current density

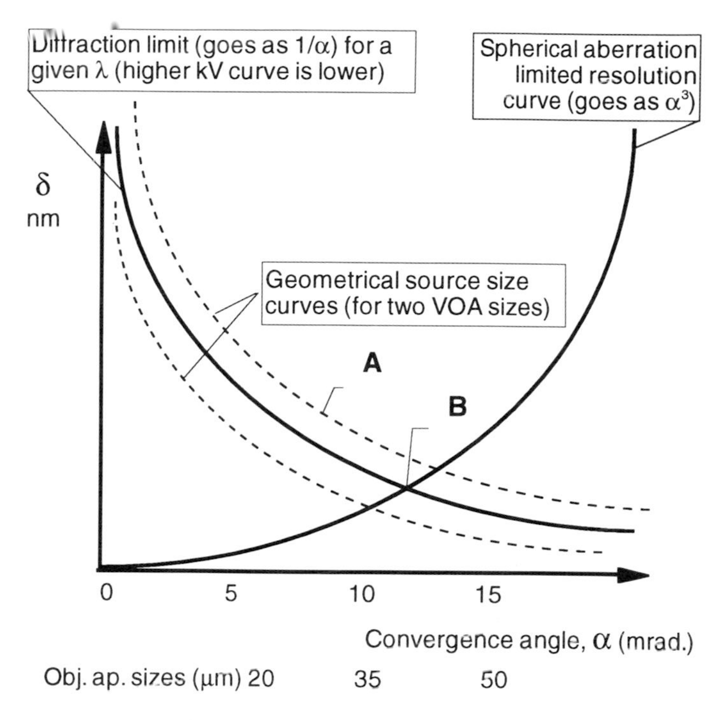

Figure 4.10. A set of graphs that shows how the geometrical probe size, δ (FWHM), is affected by convergence angle using a VOA to define the probe. Curves for diffraction and spherical aberration are also included. A few real objective aperture sizes for a VG601UX STEM are shown since changing the condenser lenses does not then change the convergence angle in STEM

distribution has extended wings or 'tails' (see *Figure 4.2*). That is, there is a significant fraction of the total beam current spread over a wide area beyond the central (Gaussian) part of the probe. This will tend to spoil our chemical signal, and may give rise to beam damage spread over a larger area than the nominal probe size. What we ought to do is focus carefully (e.g. using the ADF image) and choose a smaller convergence angle (such as A in *Figure 4.10*) since this will yield a tidier probe (more Gaussian) without the problem of the wings. One might object that the smaller angle will have less current within it, but this need not be the case if we have the virtual objective aperture in combination with two condenser lenses since then the convergence angle is continuously variable while the current is the same.

When we use a real STEM objective aperture the convergence angle at the specimen is fixed by the size of the aperture and the height of the specimen. In this case the effects of demagnification are to reduce the source size and instabilities to a level that allows the diffraction and spherical aberration limits to be reached (at point B of *Figure 4.10*). In (S)TEM the condenser aperture is the effective source-limiting aperture and it may be treated somewhat like the VOA case above for STEM.

4.5.4 Scattering inside specimens

The initial probe size is made larger inside most specimens by the scattering of the electrons through a range of angles depending on their interactions with the atoms of the specimen. Scattering of electrons outside the probe envelope (about 90% of the incident beam current is normally contained within 1.8 times the quoted FWHM probe size) mostly occurs by high energy electron interactions with atomic forces. Collective interactions with free electrons (the Fermi 'gas') gives rise to relatively small angles of scatter, although Bragg diffraction (elastic lattice scattering) provides a means for scattering outside a typical probe envelope angle (convergence angles and Bragg angles often are about 10 mrad or half a degree). A strong channelling effect occurs when crystals are oriented exactly along high symmetry axes, and beam spreading can then be much reduced.

Precisely what happens inside a crystal while electrons are passing through is open to interpretation. At the limit of small probe size the convergence angle is often larger than the Bragg angle, so that in a simple model diffracted beams form diverging cones that overlap at their edges. We may see this overlap directly with the diffraction screen (by viewing the convergent beam diffraction pattern, see Chapter 5). By sampling the overlap region of the diffraction discs with a sufficiently small collector aperture (whilst scanning in image mode), bright field lattice images may be seen. These fringe patterns behave just like TEM lattice images: there is an optimum under-focus (Scherzer defocus) and contrast reversals with focus and specimen thickness are observed.

4.5.5 Atomic number contrast

When we discard the diffracted discs by collecting only the high angle scattered electrons (in the diffraction pattern these form a diffuse background) we find that images with high resolution are also observed. These high-angle annular dark field (HAADF) images have atomic number contrast since higher atomic number atoms are more likely to scatter to high angles. Furthermore the images do not have contrast reversals with changes in focus and specimen thickness. Experiments have confirmed that HAADF images can show more detail (and more easily interpreted detail at that) than the STEM analogue to TEM BF lattice images. Backscattered electrons form images that are also strongly dependent on atomic number, since the scattering probability is strongly dependent on the number of protons in the nucleus.

4.5.6 Thickness effects

Probe or beam broadening in thick specimens degrades the point resolution of the image (or analysis) because signals may arise from a larger volume of material, with a diameter larger than the apparent beam diameter. EDXS signal intensity rises (as do X-ray absorption effects) while EELS signals become complicated by multiple scattering.

If we want to understand the effect of thickness on imaging we shall have to consider more details about detector geometries. Bright field detectors with a small acceptance angle prevent most singly scattered electrons from reaching the scintillator, image intensity drops, and thick areas look dark compared to thin areas. Multiple (or plural) scattering can result in some scattered electrons being scattered back into the BF detector, but this effect is important only if the average electron entering the specimen is scattered more than once.

Detectors that are not angle sensitive (e.g. ADF type detectors) respond equally well to electrons scattered by small angles after their initial scattering and no major differences will be seen in the image. Note that ADF images get brighter with thickness up to a limit where the total scattering angle exceeds the outer cut-off angle for the detector (typically 150 mrad). Really thick specimens look dark in both BF and ADF type images.

Penetration is a term used to describe the mass-thickness of material that yields a certain image resolution; it depends on aperture angle and accelerating potential. STEM is largely immune to the effects of chromatic aberration in post specimen optics as image resolution is related to detectable changes in signals that correspond to changes in probe position, irrespective of how tortuous the path taken by the electrons in reaching the detector. The 100 kV FEG-STEM competes well with high voltage (200–400 kV) TEMs in terms of penetration when the ADF image, or a large collector BF image, is collected.

4.6 Comparisons with CTEM

In *Figure 4.11* electrons are incident on a crystalline material that is strongly diffracting (in the case illustrated the beam is tilted by a Bragg angle). Only parallel beams travelling parallel to the lens axis are focused to a point on the axis (the principal focal point); all other parallel beams (inclined to the axis because of diffraction for instance) are focused to points (e.g. at A in *Figure 4.11*) that lie in a plane through the principal focal point in a plane perpendicular to the axis. This plane is called the back focal or diffraction plane. The TEM objective aperture lies in the diffraction plane and usually filters out some of the diffracted electrons, producing diffraction contrast in the image.

The distribution of intensity in the diffraction plane is measured in angular units that correspond to the angles of incidence of the beams **into the lens**, not just diffracted angles from the specimen. This is important because the STEM has a range of incident angles of the

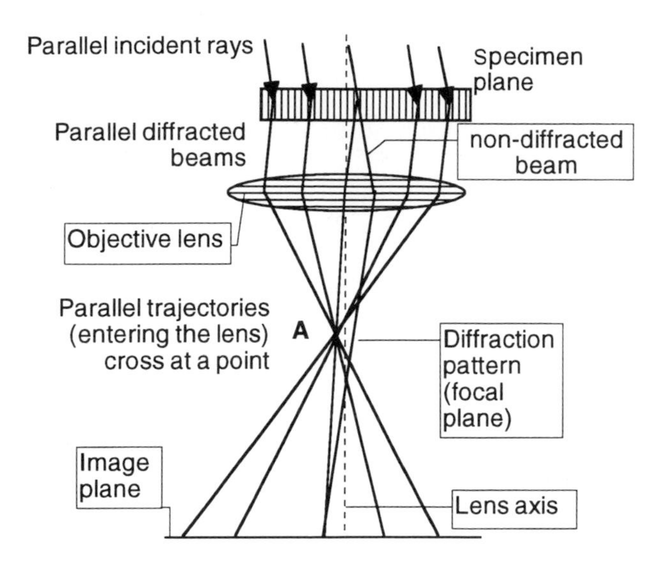

Figure 4.11. The ideal TEM imaging system. Some of the electrons incident as parallel rays on a specimen are diffracted by crystal planes. All electrons are focused to form a diffraction pattern and an image. For clarity most of the non-diffracted electrons are omitted.

converging beam on to the specimen and all these are in turn diffracted to produce discs of intensity in the focal plane, some of which may overlap (see *Figure 5.3*).

The TEM case illustrated in *Figure 4.11* needs to be generalized (as *Figure 4.12* shows) by imagining the incident, tilted illumination forming a cone of converging radiation around the incident probe direction. This cone in turn is reproduced about each of the diffracted beam directions, so that each spot in the TEM diffraction plane is replaced by a disc.

4.7 Relationships with diffraction

Conventionally the diffraction pattern is not regarded as an image, and optical ray diagrams clearly show the two to be quite separate in all microscopes. While the diffraction pattern is the electron distribution in the objective back focal plane in a transmission microscope (see *Figure 4.11*), the TEM image is the distribution of electron intensity in the image plane, where transmitted, diffracted, and scattered electrons arising from a point on the object recombine.

In a STEM no image plane is used (even if it exists). In a digitally controlled STEM the probe is moved in steps, spending a short (dwell) time after each step and parts of the stationary probe diffraction pattern

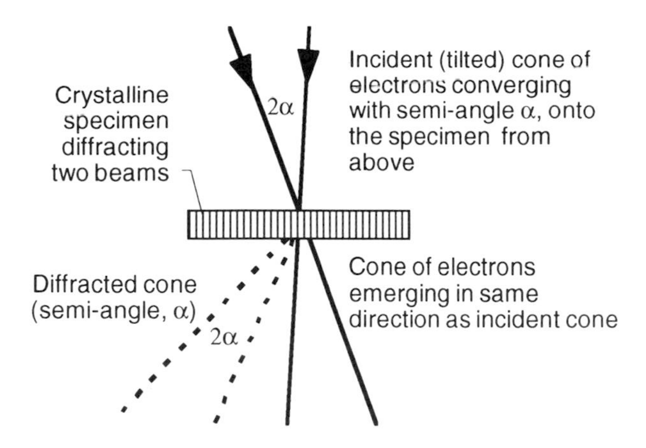

Figure 4.12. Stationary diffraction pattern formation in a STEM. The incident probe is tilted as in *Figure 4.11* for strong Bragg diffraction.

(*Figure 4.12*) are collected. After integration of the detected intensity (over pixel dwell-time and the angles shown in *Figure 4.4*) the result is displayed as a function of probe position on the specimen, and is called the **image**. The STEM integrates information contained in the diffraction pattern serially in the detector and electronics system while the TEM does the integration in parallel in the lens action of the strong post-field; notice that *Figure 4.12* is drawn (for simplicity) without showing any lens effects.

4.8 A few more numbers

Since a statistically significant number of electrons is needed to form a reasonable signal and detectors are not perfectly efficient or do not collect all available electrons, very small probes are not the best for analytical purposes. We need a current of around a nanoamp (10^{-9} A) to give enough signal for spectroscopy and although imaging electrons are much more efficiently collected we still need about 10 picoamps (10^{-11} A) to see a good image. For a typical current (say 10^{-10} A) there are about 6×10^8 electrons per second, if the dwell time per pixel is 5 μsec (giving a 512×512 frame in 1.3 sec), there are about 3000 electrons incident on each point of the specimen per frame. At TV scanning rates (50 frames per second or 60 depending on where you are working) this drops to about 100 electrons per pixel and if we collect only one frame in 100 seconds we still get 'only' about 200 000 electrons per pixel. These electrons must generate a statistically significant signal; the detection system must in turn be efficient, or the final image will be noisy (compare the noise in *Figure 4.9a* and *c*).

References

Hecht, E. (1990) *Optics*, 2nd Edn. Addison-Wesley, Reading, MA, p. 371.
Jenkins, F.A. and White, H.E. (1981) *Fundamentals of Optics*, 4th Edn, McGraw-Hill, New York, p. 327.
Longhurst, R.S. (1973) *Geometrical and Physical Optics*, 3rd Edn. Longman, London, p. 239.

5 Diffraction in the STEM

5.1 Introduction

Make a small hole in a piece of card, and you have a pinhole camera. Hold the card close to your eye and look through the aperture at a light far away (if it is a yellow street light, so much the better). You will most likely see the lamp framed in a small circle with some faint rings around it. If you place the card on to a firm surface and make the hole while holding the pin firmly, the aperture will appear much smaller. Looking through the small aperture at a sodium street lamp you will see a dimmed image of the lamp, but now with a more distinctly defined yet coarser graininess also arranged in rings.

This is the Fresnel diffraction pattern from your pinhole aperture. Diffraction patterns are also commonly observed when street lights are seen through an umbrella and when star-like patterns appear around bright headlights; also, haloes around the moon are diffraction patterns. Diffraction is unavoidable whenever scattering takes place, it is the fundamental limit to the resolution of small objects.

The most important quantity that distinguishes the different types of electron diffraction pattern is the convergence angle of the incident radiation. Since the diffraction pattern is a map of angles (where distance measured in the pattern corresponds to angles of diffraction) the highly convergent envelope, used for small probe size, does not generally provide high angular resolution in the diffraction pattern.

5.1.1 Lens effects

In a system where no image plane exists (such as a STEM with no post-specimen lenses) there may be a useful diffraction pattern; in many cases that is all there is. In a TEM a diffraction pattern is located in the back focal plane of the objective lens (parallel rays entering the lens meet at a point in the focal plane, see *Figure 4.11*). The optical terms Fraunhofer and Fresnel diffraction apply equally well to electrons; the former is the pattern seen on a distant screen while the latter is that seen close to. If the lens has no real focal plane (i.e. the lens is too weak to bring even the diffracted rays to a focus) a defocused Fraunhofer diffraction pattern is often found at a considerable distance from the lens. Post-specimen lenses are able to properly focus the diffraction pattern on to a fluorescent screen for observation and subsequent recording.

If you defocus an image formed from several diffracted beams (using a lens) you begin to see diffraction effects mixed in with the out-of-focus image. Indeed when the specimen is oriented to produce strongly diffracted beams, the image breaks up into several separate images that move apart as the focus is changed more. Only when the lens is well focused do the diffracted parts combine properly to form the image. The breaking apart of out-of-focus images under strong diffraction conditions is used as a method to measure spherical aberration, since the sideways movement of the separate images is a result of this type of aberration. Although an image is easier to relate to intuitively, the diffraction pattern is often easier to describe mathematically (see, for example, *Figure 5.4*).

5.1.2 Reciprocity

It is surprising that the bright field (BF) STEM images correspond so closely to the BF image seen in a TEM (e.g. *Figures 4.6–4.8*). In fact the correspondence between the images is only close if the range of angular integration in the BF STEM detector plane (see *Figure 4.4*) is close to the angle of beam convergence in the TEM. These two quantities are defined on opposite sides of the specimen and the correspondence is referred to as a reciprocal relationship (reciprocity). The reciprocal relationship results from considering the measure of image resolution in the TEM while in STEM the measure is the incident probe envelope. A point in the image plane of a TEM image is directly traceable to a point on the specimen exit surface; and the transfer of information is described by a function that, mathematically, has the same form as the function that describes the STEM probe on the specimen entrance surface in terms of the electron source.

In order finally to put STEM image formation aside consider *Figure 5.1* which shows how the scanning system produces at each point in the scan a different diffraction pattern that may be sampled by the electron detectors in different ways.

The mechanism of image formation is different for STEM and TEM. In TEM the objective lens combines the transmitted electrons into an image, while in STEM the detector selects some of the electrons and stores their information. Different regions of the specimen will diffract differently (in different directions and/or by different amounts). Contrast in TEM is determined mostly by the condenser aperture and the filtering effect of the objective aperture which is in the back focal/diffraction pattern plane after the specimen; while in STEM the contrast is controlled both by the objective aperture which is before the specimen

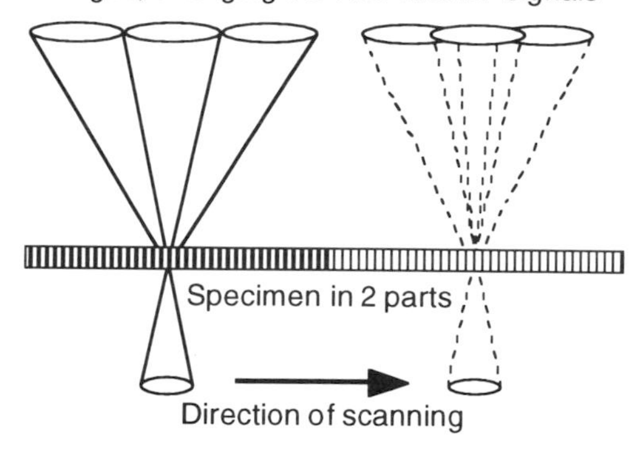

Figure 5.1. A STEM image is originated by scanning a focused beam of electrons (incident from below) across a specimen and recording, as a function of position, either a small or a large part of the diffraction pattern (like the SEM, the STEM is a truly electronic microscope). The angles of diffraction accepted by the detectors define what form the image will take (see *Figure 4.4*).

(in the front focal plane) and by the filtering effect of the detector's collection angles (which are defined in the plane of the diffraction pattern, see *Figure 4.4*).

5.2 Selected area diffraction

The repetitive nature of the atomic arrangement in crystals gives rise to Bragg diffraction patterns, and these in turn give information about crystal orientation and lattice parameter. Angular resolution is the critical issue, especially if the only measure of the diffraction is the position of the diffracted beams (spots or discs); angular resolution is highest in selected area diffraction patterns and the fine structure sometimes visible in convergent beam patterns. First we review TEM SAD before turning to the STEM version of SAD and in Section 5.3 we explore other types of diffraction.

5.2.1 *TEM image magnification*

The TEM image is formed by the objective lens (see *Figure 4.11*) and is magnified by projector lenses to provide a range of final images that, with careful adjustment, are in focus and free from distortion. Ideally the objective lens setting does not change when we magnify and only the projector lenses change; taking the primary image as their object and magnifying it into a second (then a third or even a fourth) image at higher and higher magnifications. Finally, on the phosphor screen of the TEM, we see the final image projected at a magnification given by an overall factor derived from the intermediate lenses going back to the specimen. Note the similarity here with probe formation and the magnification (or rather condenser lens demagnification) of the source in STEM.

5.2.2 Selected area diffraction in TEM

Instead of using the primary image as an object for the projector lenses we can choose to use the diffraction pattern as the object for the projector lens system. That is, we change the TEM projector lens system to transfer (and magnify) the diffraction pattern on to the viewing screen. The diffraction pattern exists in the back focal plane of the objective lens (again see *Figure 4.11*) and it is a perfectly valid object as far as the projectors are concerned.

Since the objective lens is unchanged its image of the specimen is unchanged and an aperture in this image plane, which is conjugate to the specimen plane, restricts electrons proceeding further along the train of projector lenses. Electrons that emerge from the specimen at points corresponding to points inside the aperture go on to reach the viewing screen. Thus an area on the specimen is selected by the **selected area aperture** (or field limiting aperture as it is otherwise known). How does this work in a STEM?

5.2.3 Selected area diffraction in STEM

In the simplest STEM there are no projector lenses, as we have seen in Chapter 4. The standard method of obtaining SAD patterns in STEM is to change the scanning system and the incident beam conditions to give a rocking beam pattern, just as in many SEMs. *Figure 5.2* shows a ray diagram for the simplest STEM system that uses a scanning system to rock the beam about an area on the specimen that is defined by the SAD aperture (Section 2.3.5). The resulting signal from the (axial) BF detector is the angular distribution of scattering emerging from the specimen (i.e. the diffraction pattern). How it works is described in detail below.

(i) First the SAD aperture is placed at a conjugate position to the specimen. This is set up by focusing a probe using the C2 lens on to the edge of the SAD aperture (see Section 2.8). An image of the SAD aperture appears on the monitor since the scanning system comes after C2 but before the SAD aperture.
(ii) Next, the probe is transferred on to the specimen by the objective lens with a demagnification of about 50 (so the SAD aperture and the specimen are now in conjugate planes). Thus far we have merely set up the imaging mode correctly in preparation for the transition to the diffraction mode.

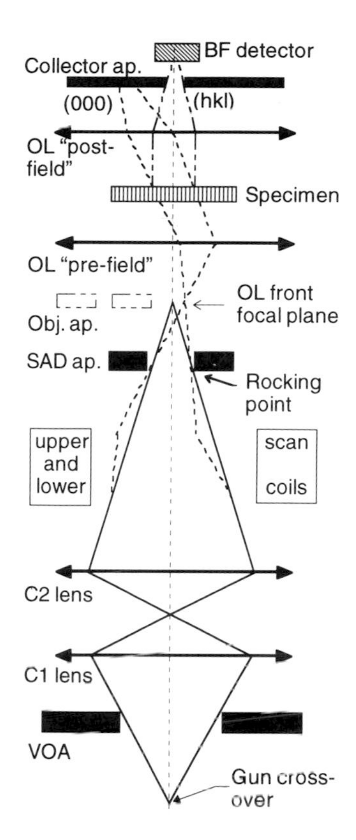

Figure 5.2. Schematic diagram showing the STEM SA diffraction mode. a ray diagram for the VG STEM HB5 series with two condenser lenses and a VOA (with C2 focused at the objective lens front focal plane). The real objective aperture is at the OL front focal plane and nominally parallel illumination falls on to the sample. Note that the source position is shown at the bottom and may be taken as that produced near the DPA by the GL. As shown, C1 is demagnifying and C2 is magnifying, but the diagram is not to scale. Compare with *Figure 2.3.* The scan coils are energized to rock the beam about the SAD aperture and hence also about the disc of illumination on the sample. Signals, falling on the axial bright field detector, map out the diffraction pattern during the scan. For simplicity only the scanned beam is shown falling on to the specimen and only two diffracted beams are shown leaving it. Notice that the electrons are brought to a focus (at about the plane of the collector aperture) in the 'far-field' behind the specimen by the 'post-field' of the OL. Since the detectors are all at about the same place in the microscope it is found that a picture is formed as the focused electrons are scanned over the detector plane (e.g. the ADF scintillator may be inspected for damage). In this way one can verify that indeed most of the diffracted electrons pass right through the HAADF detector's inner hole.

(iii) The C2 lens is refocused (relaxed) on to the front focal plane of the objective lens so that parallel illumination then falls on to the specimen (by definition of focal length, a point source at the focal point of a lens will yield parallel rays).

(iv) Finally, the scanning system is changed so that the beam is rocked about the SAD aperture through angles wide enough to include several Bragg diffraction angles. By the action of the objective lens the beam rocks on the specimen also, since the specimen and the

SAD aperture planes are conjugate. Thus a parallel electron beam is rocked through a two-dimensional set of angles on an area on the specimen.

The resulting signals received at the collector aperture/BF scintillator are displayed. Diffracted and non-diffracted beams are brought to near focus in the far field by the post-field of the objective lens. At some stage the two-dimensional scanning system will pass through zero excitation so the bright field electrons can enter the collector aperture and a bright disc (that is the collector aperture) will appear on the monitor. The diffraction pattern is scanned over the collector aperture and we see a map of angles, which is the STEM version of a SAD pattern. The 'camera length' is governed by the angular range of the scanning system and the size of discs is defined by the collector aperture angle subtended at the specimen. Examples of two SAD patterns taken with different collector aperture angles are shown in *Figure 5.3b* and *c*.

The real objective aperture will be clearly visible (as in the TEM) as a shadow image in the SAD pattern since a focused probe scans the aperture. *Figure 5.2* shows that because the rocking point has moved, the objective aperture restricts the angles of incidence to a cone defined geometrically on the specimen. The semi-angle of the cone will be the same as the convergence angle in imaging mode. Only those portions of the angular scan which are close to the lens axis (and by implication within the objective aperture) will result in detection by the axial BF detector. This is an example of the reciprocal relationship between incident angles and diffracted angles. Removing the real objective aperture will scan the full SAD pattern. The VOA would not be visible as it is no longer conjugate to the real objective aperture (cf. Section 2.2.4). Post-specimen alignments can direct particular diffracted beams into the apparent objective aperture for dark field imaging and so act as the STEM analogue to TEM beam tilt. We shall discuss the contrast implications of the objective aperture in Section 5.3.1.

The high-angle ADF detector collects the diffuse background of the diffraction pattern, which originates from inelastic scattering in the specimen. That is, the ADF display shows a Kikuchi map, without diffraction spots. This separation of diffraction spots in the BF signal and Kikuchi bands in the ADF signal is very useful when changing specimen orientation.

If we stop the scan, but still use a parallel incident beam, and view what emerges from the specimen on the diffraction screen (DPOS), a SAD pattern similar to that seen in a TEM appears. The convergence angle at the OL front focal plane is determined by the SAD aperture (see *Figure 5.2*, solid line), and it may be as small as 0.1 mrad (e.g. 20 μm aperture about 100 mm away). The selected area on the specimen is quite large, about 400 nm diameter (in the example) for a 2 mm focal length objective. The spherical aberration error associated with this selected area depends on the Bragg angle; for a Bragg angle (θ_B) of 10 mrad and a C_s of 1 mm we find, in the above example, an error of about 100 nm. Thus the spatial resolution limits on SAD in the STEM are much the same as in CTEM.

5.2.4 Post-specimen compression

Measuring the diameter of discs in the standard, rocking beam, (S)TEM SAD pattern, calibrated from a known lattice spacing, gives the angle of acceptance (2β) of the collector aperture (see *Figure 5.3b* and *c*). If the geometrical size of the aperture is known (in μm) and its physical distance from the specimen is also known, the angle calculated geometrically will be found to be somewhat less than that derived from the SAD pattern. The ratio of the acceptance angle of the collector aperture referred back to the specimen as measured in the diffraction pattern to the geometrical angle derived from the mechanical drawings is called the post-specimen compression. Typical values of post-specimen compression in VG STEMs are between 2 and 5.

Post-specimen compression is a useful concept and its value changes as a function of objective lens strength. The angles accepted by the collector aperture (and hence by an EELS spectrometer too) are variable, depending on the relative height of the specimen inside the lens. If we change the lens strength when viewing different parts of a tilted specimen, we necessarily change the post-specimen compression. To maintain the same collection angles during an experiment we ought to change the height of the specimen holder to refocus the image (after moving across a tilted specimen). In TEM it is well known that image magnification (and diffraction camera length) values are only 'correct' if the standard objective lens current is used to focus the image, normally the eucentric height value.

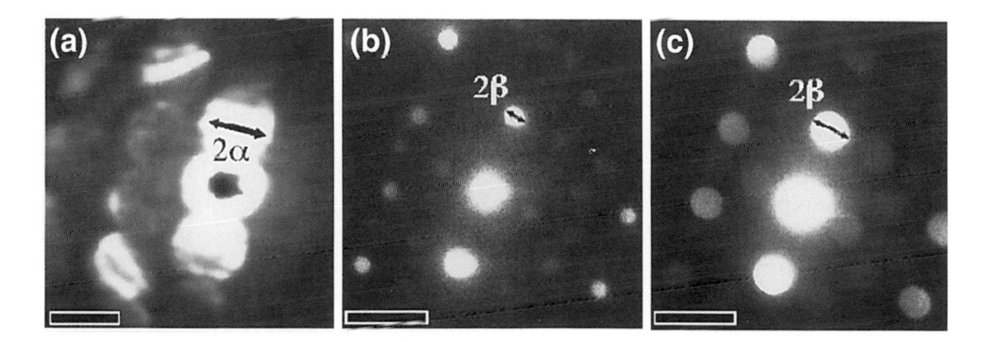

Figure 5.3. Examples of STEM diffraction patterns from the same crystal. (a) CBED (α = 7.5 mrad), (b) SAD (β = 1.7 mrad), (c) SAD (β = 3.4 mrad). Scale markers represent (a) 15 mrad, (b) 18 mrad, (c) 18 mrad.

5.3 Other types of diffraction pattern

The main types of diffraction pattern used in both TEM and STEM, apart from SAD, are of the converging and stationary beam type. The patterns are classified in terms of convergence angle relative to the diffraction angles associated with the material for the relevant wavelength. Thus if the diffraction discs are heavily overlapping, because the diffraction angles are much smaller than the convergence angle, we get a large-angle CBED or Kossel-type pattern. If the discs only overlap a little at the edges (see *Figures 5.3a* and *5.4*) we get a conventional CBED pattern; while if the discs are well resolved from each other we have a microdiffraction pattern. In the limit of very small beam convergence (i.e. when the beam is parallel) we have the normal TEM spot or ring patterns, already discussed above.

5.3.1 Convergent beam electron diffraction (CBED)

The incident probe in STEM is usually highly convergent, often as much as 10–15 mrad, in order to obtain the smallest probe diameter. By stopping the scanning and arranging a suitable imaging system to view the intensity falling on a screen far away from the specimen (with a 35 mm camera perhaps), a Fraunhofer diffraction pattern is visible. The pattern is usually found to be of very good quality despite there being no lenses other than the weak post-specimen field of the objective lens. The appearance of the STEM CBED pattern is a conventional set of overlapping discs, the diameter of which is defined by the probe-forming aperture angle (see *Figure 5.4*). Bright field STEM lattice images are only possible if the discs do overlap (for more details see Section 5.4).

The convergence angle is defined by the objective aperture where that is the probe forming aperture (i.e. in STEM) or by the condenser aperture in a (S)TEM. If the beam has to traverse any condenser lenses after the probe is defined then the convergence angle will be changed from its geometrically determined size. From the CBED pattern it is possible to measure the probe convergence angle as long as the camera length is known, or the pattern contains diffracted discs arising from known atomic plane spacings. The spacing of the disc centres in the CBED pattern is twice the Bragg angle for the lattice planes concerned and the diameter of each disc is twice the convergence angle, as *Figure 5.4* shows.

Figure 5.5 is a CBED pattern (from a quantum well structure) where the convergence angle just about matches the Bragg angle perfectly (the discs touch but hardly overlap). The picture is taken from a VG STEM (fitted with a slow-scan CCD system) and shows that even without the benefit of post-specimen lenses, high quality CBED information is available. The 18-bit resolution allows 'variable exposure', into three portions of 6 bits each, of the different parts of the pattern, overcoming

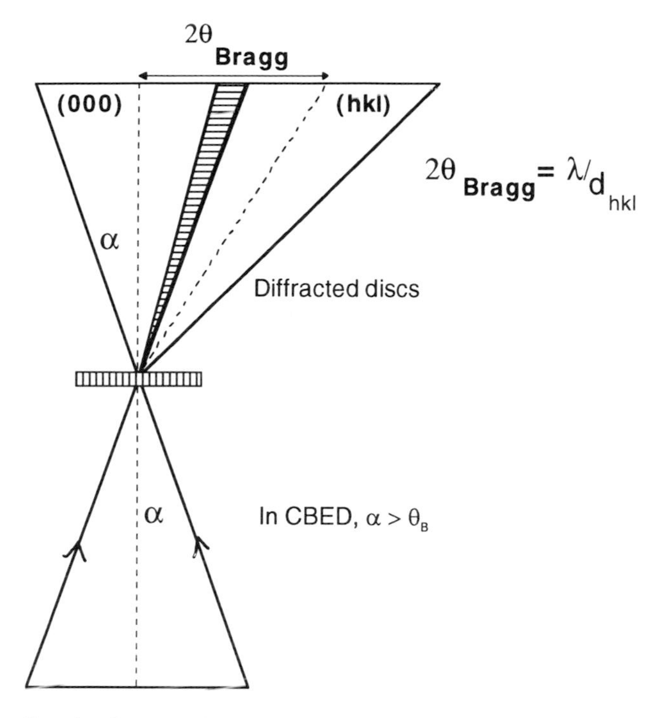

Figure 5.4. Simple diagram showing the relationship between the angle of convergence (defined by the beam defining aperture) and the angles of diffraction, which separates microdiffraction from convergent beam diffraction. Two diffracted discs are obtained: the (000) 'bright-field' beam and the beam diffracted from planes with Miller indices (*hkl*). The spacing of the diffracting crystal planes is d_{hkl}. In high resolution STEM imaging a large convergence angle is necessary. A convergent beam diffraction pattern is formed when the convergence angle, α, is greater than the Bragg angle, θ_B. The overlap of the diffracted discs leads to interference effects, such as lattice fringes.

the problem that the central parts of diffraction patterns are usually very much the most intense. This example effectively answers any criticism regarding the performance of the VG STEM in producing good quality diffraction patterns.

Information can be obtained concerning three-dimensional crystal symmetry from the symmetry of the diffracted intensity in the whole pattern and from within the central CBED discs. The fine dark lines in the central disc are known as deficit Kikuchi lines and correspond precisely with bright (excess) lines that appear in the large angle ring of discs surrounding the central part of the CBED pattern. The high angle ring of discs originates from the excitation of diffracted intensity from crystal planes along the probe axis. The fine dark lines are very sensitive to changes in local lattice spacing and comparison with simulations allows very accurate lattice parameter measurements; the relative positions of the fine lines in the central disc of *Figure 5.5* were used to determine the state of strain in the specimen.

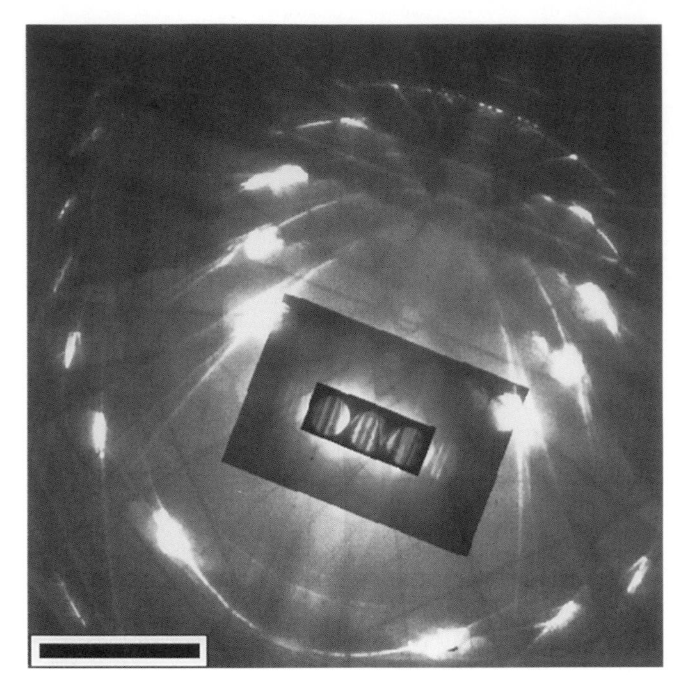

Figure 5.5. A CBED pattern of gallium nitride taken with a VG STEM; the information is used for strain measurements. Note that the slow-scan CCD camera provides a high quality output and allows 'variable exposures' to improve dynamic range. Bar = 55 mrad. The full field of view in the pattern is 220 mrad (i.e. about 12.6°). (Courtesy Dr H. Lakner, Duisburg University.)

5.3.2 Microdiffraction

If the beam convergence angle is smaller than the Bragg angle (see *Figure 5.4*) the diffraction pattern obtained is sometimes called a microdiffraction pattern. Numerous attempts have been made to form a small diameter near-parallel beam of electrons, which would allow high angular resolution diffraction information to be recorded. High angular resolution is only in general possible by sacrificing spatial resolution, as we saw in an example at the end of Section 5.2.3. Some (S)TEM instruments make use of a third condenser lens to allow special modes of operation that can provide relatively narrow (and near parallel) probes to be formed. A beam diameter of about 20 nm is possible and although the current is not very great, good quality diffraction patterns may be recorded. Since these are non-scanning techniques we shall not pursue them further.

A different approach to recording the diffraction pattern from a small probed area (also called microdiffraction) is illustrated in *Figure 5.6*. The diffraction pattern from a static probe is scanned across the collector aperture (so-called Grigson scanning), and the result is a sampling of the pattern. Furthermore, the scan could inject the pattern into an imaging EELS system, thus filtering the inelastically scattered background. The

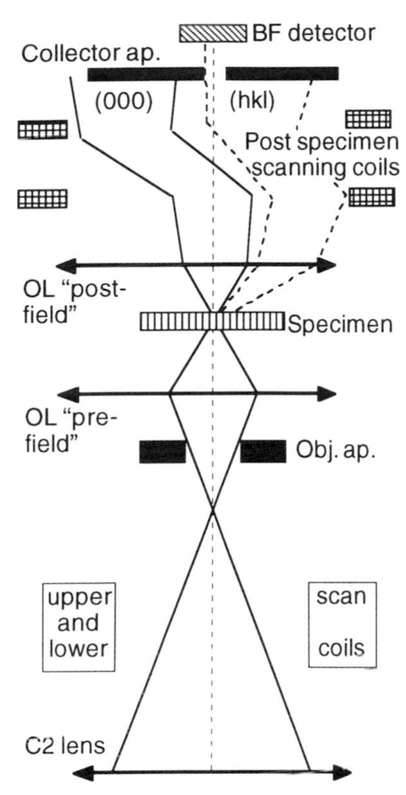

Figure 5.6. Microdiffraction: the incident beam is focused and stationary on the specimen (in spot mode). The CBED pattern is scanned over the (small) collector aperture achieving a high resolution sampling of the pattern; it also allows energy filtering if the BF detector is after an electron energy-analyser.

method is strongly related to a technique called 'de-scanning' which finds utility in low magnification energy-filtered imaging, where highly displaced electrons otherwise do not pass through the collector and a cut-off in the image is visible.

5.4 High resolution STEM imaging

Although logically one would expect to find this section in Chapter 4 (see Section 4.5.4) it appears here because it draws heavily on some of the preceding discussion.

The objective aperture has a filtering effect in the BF imaging mode in TEM. In STEM it controls the convergence angle and this controls the contrast. If, for example, we arrange for several beams to appear inside the objective aperture in the SAD pattern, we might hope to obtain lattice images; if we view the DPOS output we shall verify that the CBED discs overlap. If in the SAD pattern the objective aperture is too small to include diffracted beams, then no BF lattice images could appear; in stationary probe STEM image mode we note that the DPOS shows a microdiffraction pattern (the discs do not overlap). Furthermore if the convergence angle is too small to cause CBED then the diffraction limit (see Section 4.2.1 and *Figure 4.10*) limits the probe size, so that HAADF images do not show atomic column resolution either. In this example, we cannot see lattice resolution in BF because $\alpha < \theta_B$ (no overlap means no interference) and in HAADF the lattice spacing is smaller than the probe size, $d_{hkl} < 0.61\ \lambda/\alpha$ (recalling Bragg's law from *Figure 5.4* to relate λ, θ_B, and d_{hkl}). Setting $\alpha < \theta_B$ the reader may show that this reinforces the inequality $0.5/\theta_B < 0.61/\alpha$ and the spacing d_{hkl} is not resolved within the diffraction limit on probe size.

A unique feature of the standard VG STEM system is that it is quite easy to obtain a simultaneous representation of the image and the CBED pattern in 'real time'. The way this works is that the diffraction screen may have a hole in it (see *Figure 5.3a*), through which electrons in the overlapping CBED discs may proceed to be collected by the bright field detector. Scanning the beam across the specimen at a high magnification means the diffraction pattern will most likely not be changing much (since only small regions of the specimen are being sampled). Under the right probe size, specimen orientation, and collector aperture conditions then lattice images may be viewed whilst observing the local diffraction pattern, making the selection of optimum conditions easier.

Figure 5.7 shows a schematic representation of how moving the electron probe to different specimen positions can affect the details of the CBED pattern. In the case of a crystalline material where CBED discs overlap interference effects occur (like the Young's slits experiments using laser beams). As the probe moves from position 1 *in Figure 5.7* the fringes shift sideways slightly to that shown at position 2. This shift is the result of the slightly different diffraction conditions at the two locations (a similar shift occurs with objective lens focus). Sampling the intensity within a small BF STEM collector aperture in the overlap region of the CBED pattern, and displaying the variations that result during probe movement, allows lattice images to be recorded. The contrast is weakened when a large collector aperture is used since the shifting of the fringes is less significant compared to the collection angle. The highly coherent electron probe may cover more than one unit cell and still good contrast fringes will be visible; thus, large convergence angles still yield high-resolution BF images. HAADF images, on the other hand, need a small probe size to make the variation in the diffuse scattering that results from localized channelling of electrons along or between atom columns visible. BF lattice images are visible over a large range of focus values and contrast reversals may occur due to the fringe shift noted above; HAADF images are only visible over a small range of focus and no contrast reversals are seen. Unfortunately small collector apertures also give noisy images, and so STEM does not compete directly with TEM in the conventional BF high resolution mode.

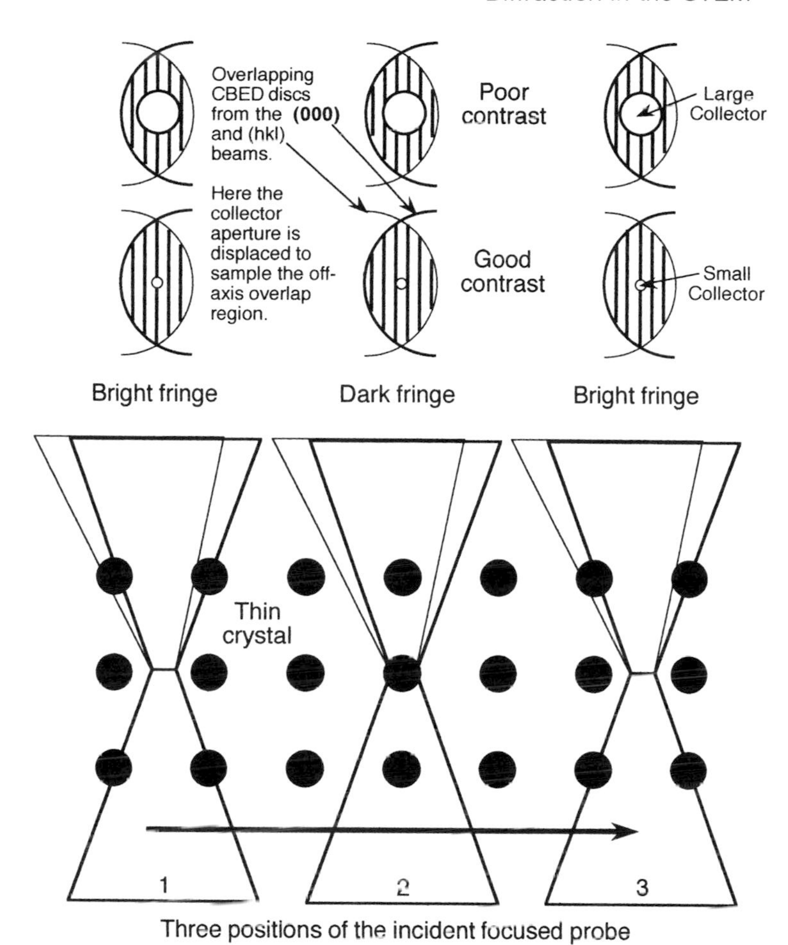

Figure 5.7. A schematic representation of the relationship between bright field lattice fringe contrast and STEM collector aperture size. In STEM the variations in intensity detected by the BF detector are displayed as the probe moves from place to place across the specimen. A small collector aperture is better able to resolve the interference fringes that appear in the overlap region of the CBED discs. High resolution BF images are possible with very large convergence angles, even though spherical aberration limits the probe size to greater than the lattice spacing. Bragg diffraction takes place across several unit cells and the FEG probe has a high degree of lateral coherence.

5.5 Limits to diffraction

The smallest area that can be probed in electron microscopy is about 0.2 nm in diameter using the type of technology we have at present. There is little point in making the area from which we derive diffraction information much smaller than 0.2 nm diameter. This area is of atomic dimensions. We can probe inside the unit cell of many materials and form diffraction patterns. Diffraction patterns from inside the unit cell do not exactly match those of the larger lattice because of breakdowns of structure factor effects (i.e. repetitions of scattering across several unit cells causing interference effects).

The ultimate limit of wavelength is not likely to be a limiting factor since the electron has a short wavelength compared to atomic dimensions. Our limit is in controlling the electrons we have. The factors that affect this control are the convergence angle of the probe envelope, the electron energy, and microscope lens aberrations. Other limiting factors include instabilities, astigmatism, interference, vibration, and beam damage.

Thin specimens often give rise to what are called kinematic diffraction patterns. Basically, this means that the average electron is only scattered (diffracted) once in passing through the specimen. However, in thicker specimens, dynamical diffraction effects give rise to extra details within the diffracted discs, such as sets of relatively broad, bright fringes. These fringes are extensively used in TEM instruments to obtain specimen thickness measurements. The same applies to STEM, especially if good quality patterns (such as *Figure 5.5*) are obtainable.

5.6 Summary

To summarize this chapter, recall how STEM uses a focused probe that has a large angle of convergence, often as large as (or larger than) the characteristic Bragg diffraction angle itself. The STEM method to produce SAD patterns makes the technique accessible to those used to using TEM. The CBED mode, where the pattern is viewed via an optical coupling to a plate camera or a TV camera remains the technique with which diffraction information at the smallest scale is obtainable. Finally, energy filtering of images and diffraction patterns is possible in STEM only if the BF detector comes after the energy loss spectrometer.

6 Microanalysis in the STEM

6.1 Introduction

Since traditionally the principal application of STEM has been (and remains) chemical microanalysis, we discuss in this chapter the main methods by which it is achieved.

We use a model of the atomic electronic structure that will be familiar to all our readers; it is the 'shell' model, where the most tightly bound atomic electrons occupy the innermost energy levels (or 'shells'). Quantum mechanics forbids an energy transfer which is less than the difference between the bound energy level and the next highest vacant energy level. One result of the interaction between an energetic (incident) electron and an atom can be the ionization of the atom or the creation of an inner-shell vacancy in the electronic structure. Hence, the energy spectrum of the energetic electrons leaving the specimen carries a signature of the electronic structure, and hence, amongst other things, the chemistry of the sample. Detection and study of this spectrum is termed 'electron energy-loss spectroscopy', and is usually referred to by the acronym 'EELS'.

If the vacancy created in the electronic structure occurs in the innermost levels (the 'core' levels) of the electron cloud, the result is an excited atom or ion, which can lose energy in a number of ways. The process of interest to us is where one of the outer shell electrons jumps into the vacant inner energy level, losing the excess energy by creating a characteristic X-ray photon, whose energy is unique to the chemistry of the excited atom or ion. The most common spectroscopy of these X-rays, in an electron microscope, is termed 'energy dispersive X-ray spectroscopy', and is given the acronym 'EDXS'.

We will first discuss EDXS in some detail, but shall return to energy loss spectroscopy in Section 6.6.

6.2 Energy dispersive X-ray microanalysis in the STEM

EDXS has been used for many years because inexpensive detectors and analysers have been available. EDXS is also used in scanning electron microscopes (SEMs), electron microprobes, and transmission electron microscopes (TEMs). Data interpretation is often relatively straightforward.

In its simplest terms, a description of EDXS analysis in the STEM might be as follows: a thin section of the specimen of interest is irradiated by a fine probe of electrons. The interaction of the electrons with the atoms of the specimen results in the generation of X-rays whose energy is characteristic of the atoms which emitted them. The X-rays are detected by a solid-state detector and analyser, producing an energy spectrum from which can be deduced the composition of the irradiated volume. The mass of material analysed is determined by the probe diameter and the thickness and density of the specimen.

For some applications, the foregoing description provides all the insight required for successful application of EDXS analysis. For most purposes almost every part requires some amplification or qualification. We note also that microanalysis in the STEM is fundamentally the same as microanalysis in the TEM. Indeed, modern TEMs are achieving a level of microanalytical performance that until now had been the sole purview of the dedicated STEM. The following discussion applies to either type of instrument.

6.2.1 EDXS detectors

We will begin at what might seem like the end, the X-ray detector. The type of detector used ubiquitously on electron microscopes is essentially a reverse-biased semiconductor diode. It is usually made of single crystal silicon, although detectors made of germanium are also available and have some advantages and disadvantages. It is beyond the scope of this book to describe the details of the operation of the detector, other than the observation that X-rays give rise to 'pulses' that are analysed electronically. Suffice it to say that the crystal must be kept at a temperature close to that of liquid nitrogen, and for good performance it is critical that the crystal surface be kept absolutely clean. For this reason (and, indeed, for others) the detector must be kept in a high quality, clean vacuum at all times. This is most easily done by building the detector into its own vacuum system, pumped by molecular sieve material cooled by the liquid nitrogen reservoir.

Figure 6.1 shows a greatly simplified sketch of the essentials, illustrating the crystal on the end of a copper rod (for heat conduction)

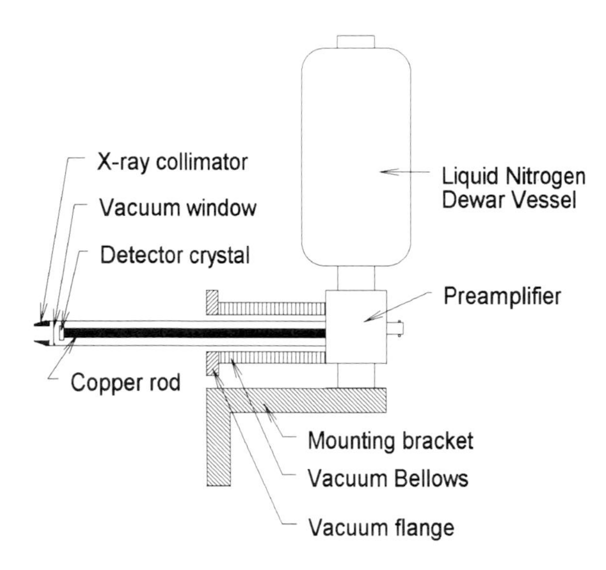

Figure 6.1. Simplified sketch of the essentials of an energy dispersive X-ray detector as used on a STEM.

enclosed within an outer tube, closed at the end by an X-ray transparent window (see the next section). Also shown is a collimator which helps to limit the detection of stray X-rays by preventing them from reaching the crystal.

6.2.2 X-ray detector windows

Since the X-rays do not travel far in most materials, it is necessary to make special provision for them to enter the detector, through what is commonly called the 'window'. Older detectors use a thin foil of beryllium metal, which was strong enough to withstand the atmospheric pressure and transmitted most X-ray photons with an energy above about 1 keV. The only elements that do not emit characteristic X-rays with an energy higher than this are those from hydrogen to sodium.

Beryllium has a number of undesirable properties, however, so since the late 1980s the majority of detectors have been fitted with alternative windows, usually made of proprietary materials developed by the various manufacturers, termed 'thin windows'. These windows typically transmit around 20–50% of the characteristic carbon X-rays which have an energy of 284 eV, so they allow the observation of a wider range of elements than Be-window detectors. Additionally, the material of these thin windows is not toxic (unlike beryllium), and modern windows are stronger and more vacuum 'tight'. On the negative side, most thin window materials transmit light to a greater or lesser extent, which is significant because the detector crystal is also sensitive to light. While in the majority of AEM applications this is not a problem, the operator needs to be aware of the potential for difficulty, for some materials (mainly ceramics) produce copious amounts of light when bombarded by electrons. (This effect is called cathodoluminescence, and can be a useful analytical tool in some circumstances; however, it requires special detection systems that we shall not discuss further.) Symptoms of light interference include a high dead-time (see Section 6.5.5), very few X-rays recorded in the spectrum, and degraded energy resolution (see Section 6.5.4). While other problems could cause these symptoms too, a sure diagnostic is that the system returns to normal when analysing a different specimen.

6.2.3 *Windowless detectors*

In a few microscopes (mainly dedicated STEMs) the column vacuum is sufficiently good that the X-ray detector will function well without a window. Such detectors are termed 'windowless', and provide the ultimate sensitivity for light element analysis. Modern, low noise examples can clearly produce a peak from beryllium ($Z = 4$). A serendipitous advantage of windowless detectors results from their mechanical compactness, which allows them to be moved close to the specimen, so a larger number of X-rays can be detected. There are a number of problems with windowless detectors, however, ranging from the obvious necessity of providing a slide, bellows, and valve assembly to allow the detector to be retracted and isolated during unavoidable venting of the column, through the susceptibility of the detector to accidental or gradual contamination, its sensitivity to light, and the difficulty of performing quantitative analysis of the light elements (see Section 6.5.3) to subtle effects related to resolution and throughput. As a result, some researchers have chosen not to use them, even on microscopes with an excellent vacuum – indeed, some even cite the degradation of the microscope vacuum itself as a reason to avoid the windowless detector!

Figure 6.2 shows a photograph of a windowless detector fitted to a VG HB603 microscope. The greatly increased complexity, compared with the schematic of *Figure 6.1*, is quite evident.

6.3 EDXS spectrum details

The details of operation of an EDXS detector are fascinating, but beyond the scope of this book, so let us proceed by examining a typical EDXS spectrum, for example *Figure 6.3*, which is a spectrum obtained from stainless steel in a VG HB601 microscope operating at 100 keV. The display shows a histogram of the numbers of X-rays which have been detected, using bins, or 'channels' which are typically 20 eV wide, but may be other widths – the most frequently used alternative is 10 eV per

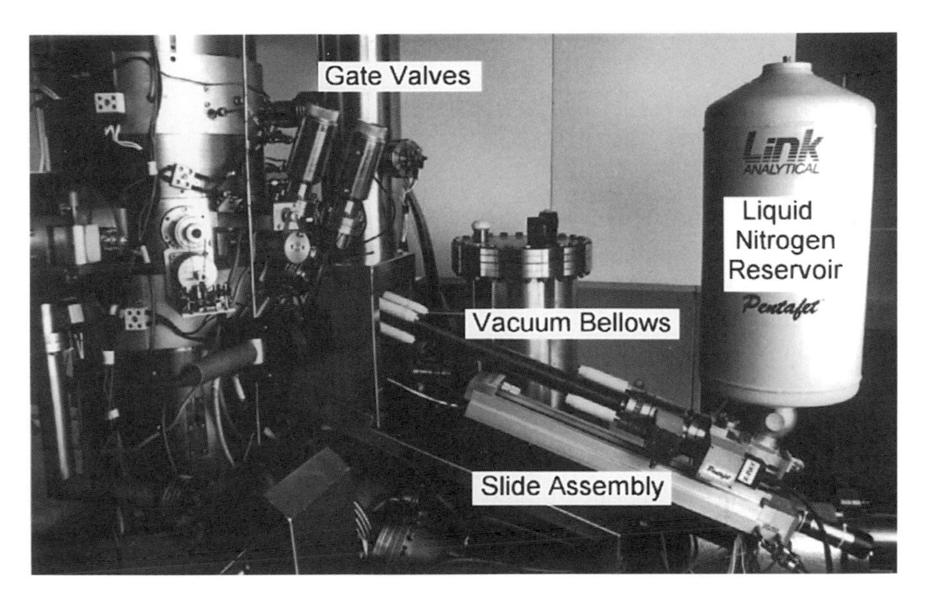

Figure 6.2. Photograph of a windowless detector fitted to the HB603 STEM at the Massachusetts Institute of Technology. The long bellows and slide assembly, and double gate-valves, allowing the detector to be retracted and isolated from the column, are indicated. Compare the complexity of this arrangement with the simple system sketched in *Figure 6.1*.

channel. Typically there are 1024 channels, leading to a spectrum width of 20.48 keV. These typical general-purpose choices can be explained by pointing out that all elements emit characteristic X-rays potentially useful for microanalysis at some energy or energies below 20 keV. In some manufacturers' systems, the zero of energy corresponds to a channel above the first one – in *Figure 6.3*, for example, zero energy is recorded in the tenth channel (although channels 1–9 are not included in the display). This is done to allow the 'zero energy' peak to be displayed; this is a useful diagnostic tool.

The spectrum itself consists of a number of peaks superimposed on a slowly varying continuum background. Although the natural energy spread of the X-rays in a particular emission line is quite small (a fraction of 1 eV) the X-ray line is recorded as a Gaussian distribution several tens of eV wide because of noise in the detector and amplifier system. It is usual to specify the resolution of the detector and analyser system at the energy of the manganese K_α radiation (the resolution varies with X-ray energy); the detector used to record *Figure 6.3* has a resolution of about 135 eV.

The background, or bremsstrahlung, is always generated when energetic electrons strike material. Besides the various peaks due directly to emission from the specimen, there may be artefacts visible in the spectrum. At least three of these result in the appearance of spurious peaks, which could result in the inexperienced analyst incorrectly deducing the presence of elements which are in fact absent, so it is quite important

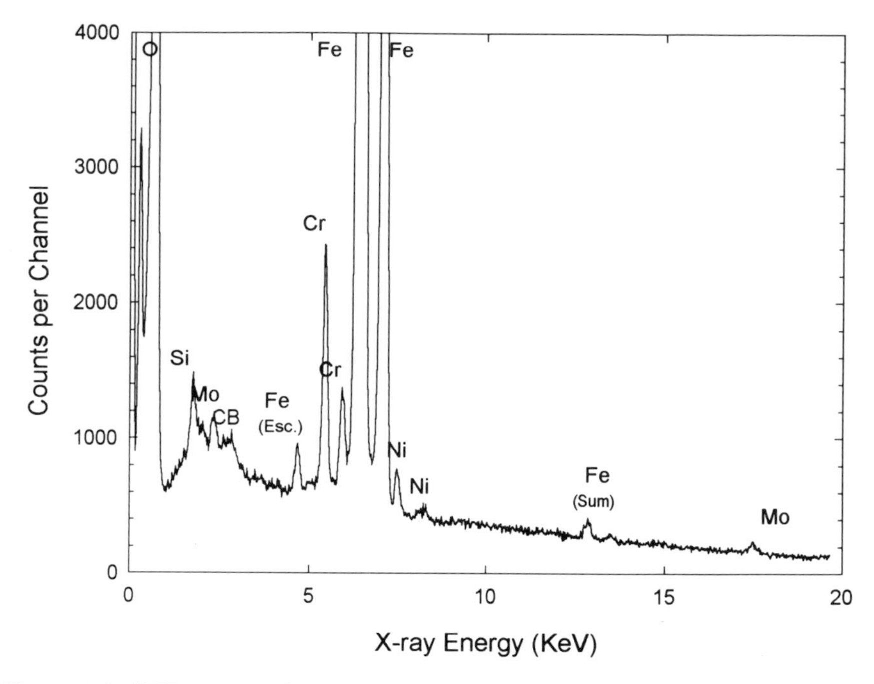

Figure 6.3. EDX spectrum from a stainless steel recorded with a windowless detector on a STEM operating at 100 kV. The characteristic peaks, as well as the escape peak from iron, the iron sum peak, and a coherent bremsstrahlung peak are labelled. This spectrum is typical of ones recorded from thicker samples; the bremsstrahlung continuum is attenuated at lower energies by self-absorption in the sample. In a spectrum from a thinner sample, the bremsstrahlung intensity would have continued to increase at lower energies.

that they be considered. In descending order of significance, these are escape peaks, sum peaks, and coherent bremsstrahlung. All three of these artefacts are visible in *Figure 6.3*. Let us describe more.

6.3.1 Escape peaks

The escape peak is a satellite peak which occurs 1.74 keV (the energy of a silicon X-ray) lower in energy than the parent peak. The peak is given its name because it results from the escape of a silicon X-ray from the detector during the detection process. For a given X-ray detector geometry the probability of generating an escape peak is well defined and depends only on the energy of the incident X-ray. All commercial X-ray analyser programs include routines for identifying and/or correcting for escape peaks. The ratio of the escape peak to the parent is constant for all specimens, and in most cases the escape peaks are the largest of the three artefacts. However, since the correction routines are usually quite precise, the escape peaks should not normally be a problem for the analyst.

6.3.2 Sum peaks

Although usually less intense, sum peaks are more simply explained. It is possible for the arrival times of two X-rays at the detector to be so close that the electronic circuitry does not detect that there are two. In this case, the result will be the apparent detection of an X-ray with an energy equal to the sum of the two incident X-rays. The probability of generating a sum peak depends on the count rate, and the largest sum peak will always occur at twice the energy of the largest parent peak in the spectrum. With modern electronics, the sum peaks only become significant at very high count rates (several thousand per second), although the coincidence detection is less efficient for lower energy X-rays (below about 1 keV). They are rarely encountered during high-resolution analysis, which, as we shall see, requires thin specimens and small electron probes, both of which result in the generation of fewer X-rays.

6.3.3 Coherent bremsstrahlung (CB)

The analyst searching for very low (<< 1%) concentrations of impurities must be aware of the possibility of observing coherent bremsstrahlung peaks. These come about from the interaction of the incident electrons with the periodic structure of crystalline specimens, and typically produce small peaks in the 1–4 keV region of the spectrum. The energy of the coherent bremsstrahlung peaks varies with the specimen orientation, the atomic spacing, and with the energy of the incident electrons.

6.4 Quantitative X-ray microanalysis

6.4.1 Cliff–Lorimer thin film method

In principle, quantitative analysis using X-rays in the STEM is quite straightforward; in practice, however, it is fraught with pitfalls. The principle is expressed in the equation, due to Cliff and Lorimer, relating the numbers N_A and N_B of X-ray counts detected from elements A and B respectively to the concentration (conventionally expressed in weight fraction) C_A and C_B of the elements in the specimen:

$$\frac{C_A}{C_B} = k_{BA}\frac{N_A}{N_B}$$

where k_{BA}, known as the 'Cliff–Lorimer k factor' is a proportionality constant (also called the 'relative sensitivity factor') which depends upon the elements being analysed, the energy of the incident electrons, and the relative sensitivity of the X-ray detector for the different X-rays. We note that this factor does not depend upon the chemical (or physical) nature of the sample. This relation is strictly valid only for infinitesimally thin specimens, but many real specimens are found to be sufficiently thin that the equation is a good approximation. (We will discuss problems related to specimen thickness shortly.) The k factor can be estimated from theoretical considerations or it can be measured by analysing a specimen of known composition. Although, for a variety of reasons, this latter technique is not as easy as it sounds, it is still the preferred method of determining the relative sensitivity factor if it is possible. If there are more than two elements present in a specimen, similar equations relate the other components, and, assuming X-rays from all elements are detected, the total composition can be found.

6.4.2 Hall method

While the foregoing discussion is quite general, there exist many situations in which it is not practical (or possible) to rely on the relationship between the characteristic peaks. The most notable example is the analysis of biological tissue, where the principal elements are all light. In the days when beryllium-window detectors were common, the light elements were quite invisible to the detector; even today, as noted in Section 6.5.1, light element detectors, while capable of detecting X-rays from these elements, have difficulties with quantitative analysis.

An alternative strategy for such investigations is due to Hall. This relies on the observed fact that within limits, the intensity of the generated bremsstrahlung is a function of the mass–thickness (the product of the sample density and the thickness) of the specimen. Hence, the ratio of the intensity of a characteristic peak from a trace element to the bremsstrahlung intensity leads, after calibration, to the desired analysis. This method relies on the microscope being free from spurious X-ray production (for that would increase the background signal), a condition that, unfortunately, cannot be taken for granted.

6.4.3 X-ray absorption

In a specimen of finite thickness, X-rays are generated throughout the depth of the foil. In the case of the simple geometry illustrated in *Figure 6.4*, an X-ray generated at depth d in the foil must travel through a distance $d/\cos(\theta)$ of the specimen, where θ is the X-ray take-off angle. While traversing this path, the X-ray may be absorbed by the material of the specimen. The problem arises because X-rays of different energy have different probabilities of absorption (very roughly, those with lower energy are absorbed more readily, but at certain energies, depending upon the material of the absorber, the absorption probability increases sharply with photon energy).

Hence, the spectrum of X-rays recorded by the detector is different from that generated in the specimen. If the exact specimen and detector geometry is known, it is possible to derive an analytic expression to

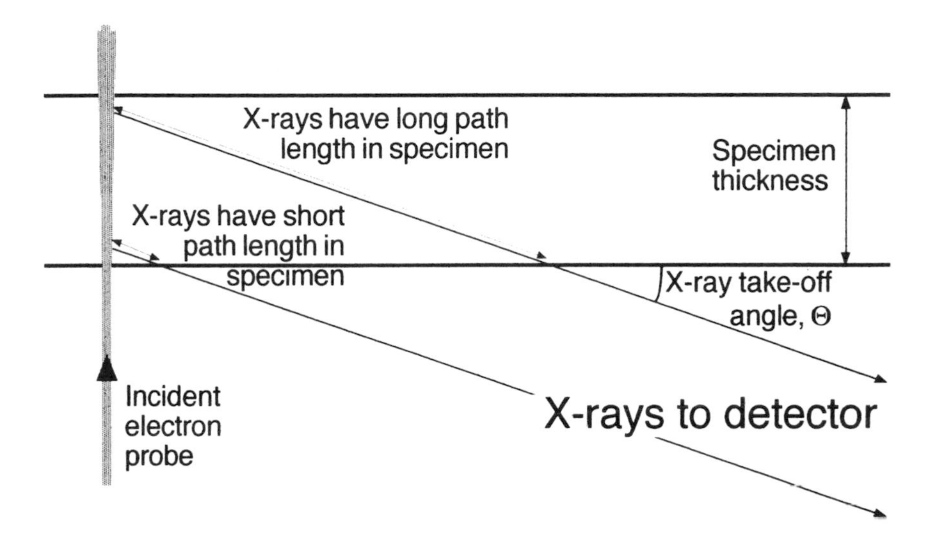

Figure 6.4. Schematic of the path of X-rays leaving a sample, illustrating the different path lengths travelled by the X-rays produced at different depths.

express (and, therefore, correct for) the absorption, but in general the specimen is not a parallel-sided slab, its tilt is known only very roughly (most specimens are buckled to some extent in their thin areas), and the specimen thickness is extremely difficult to determine. There has been some success at correcting for self-absorption by using information derived from comparing the measured background shape with theoretical models, but few, if any, commercial X-ray analysis software packages include routines for this method. Again, we do not have space to go into more detail, but it is clear that for the best quality quantitative analysis, a thin specimen is required, but the specimen must be thick enough so that adequate counts are collected (see the next section), and so that surface layers do not dominate the analysis.

6.5 Limits of EDXS analysis

In any sort of analysis, there are two different limits on the detectability of an element. The first is called the 'minimum mass fraction' – the minimum concentration of an element that can be detected when it exists combined with other elements in a specimen. The second is the 'minimum detectable mass', which is the smallest amount of the pure element that can be detected in the absence of any signal from other elements.

The principal limiting factor in the precision of X-ray microanalysis is statistics. The X-ray photons arrive randomly at the detector; such a process is described by Poisson statistics, but if the number of counts is sufficiently large (in practice, more than about 25) the use of the more well-known Gaussian statistics is an adequate approximation. Modern X-ray detectors have an upper count rate of several thousand counts per second; hence, in, let us say, 100 seconds,, a few hundred thousand X-rays can be acquired. At most about half of these would be in a major peak (the rest would be in the bremsstrahlung). We can work through the figures and show that the smallest concentration of a trace element that would be detectable in the optimum situation would be of the order of 0.1 wt%. This could be improved to 0.01 wt% by collecting the signal for 10 000 seconds (over 3 h) live time, or about 5 h when allowing for detector dead-time – about four measurements per 24-hour day! In an extreme situation such an experiment can be (and sometimes is) performed, but it cannot be considered a 'typical' analysis. There is always the possibility that the specimen may be degraded by the extraordinary electron flux it receives in this time, that contamination may build up on it, or that some other problem will prevent sufficiently accurate analysis of the spectrum (we have considered only statistical arguments in arriving at this limit).

If an experiment could be devised to demonstrate it, theory predicts that a single atom of an element such as iron, on a thin carbon substrate, would produce an easily detectable X-ray signal in 60 seconds in a modern FEG-STEM. The problem is that such an atom does not produce a detectable image – there is, therefore, no way of finding it to put the electron beam on it! A number of experiments, though, have demonstrated the detection of a few tens of atoms, both in clusters on substrates, and at boundaries. Interestingly, technology is almost at the point where the precision of the analysis is determined by the statistics of the random distribution of atoms in the sample, rather than the statistics of the counting of the X-rays! Thankfully for the microscopist, most experiments do not approach these limits!

A thin specimen must be used if high spatial resolution microanalysis is to be attempted, because the electron beam will spread laterally as it traverses the specimen. (The amount of broadening will be less for higher voltage electrons, and will be more for specimens of heavier materials.) As the specimen gets thinner, fewer X-rays are produced, degrading the statistical precision (and hence the sensitivity) of the analysis. Likewise, if the electron probe is made smaller (which can be done, within limits, by adjusting the lenses), then the current will be reduced, again leading to a loss of sensitivity.

Spatial resolution of microanalysis is a complex topic; in the extremes, for example, we may need to determine the composition of a small region of the sample to the highest precision, we may need to determine simply whether a feature is composed of one or another phase of significantly different compositions, or we may be trying to determine the amount of a trace element segregated at an interface, as in the example in Chapter 1. All these are 'high resolution microanalysis', but the way the microscope is operated and the data are acquired will be different for each. Suffice it to say that we can make a sensitive analysis of about 1000 nm^3, we can easily find the major elements in 1 nm^3, and we can detect about 10^{18} atoms m^{-2} segregated on an interface, provided that the specimen can be suitably imaged. In reality, we are often working somewhere in between these limits.

6.5.1 Light element analysis

Elements of atomic number less than 11 (sodium) are traditionally termed 'light' elements because they are not detected with systems fitted with beryllium windows. Detectors that either have no window or have a thin window, made of a material that transmits low energy X-rays efficiently, allow us to make use of these X-rays (see Sections 6.2.2 and 6.2.3). The highest energy of the emitted X-rays, from the light elements, is about 1 keV; and this results in high rates of absorption along their path to the detector crystal. The path taken is in the material of the specimen, in the window (if any), and in any layers of material on the crystal itself. Such layers include the gold contact layer, which is inherent in the design, and layers of condensates, principally ice, which inevitably form on the cold crystal, and which are of unpredictable thickness and therefore absorb unpredictable fractions of the incident X-rays. The ability to record these X-rays has extended the utility of EDXS into the realm that until the mid-1980s was the sole preserve of energy loss spectrometers. However, quantitative light-element analysis by EDXS is still difficult, and this is one area of investigation which is probably best approached by PEELS study.

6.5.2 Energy resolution

Opening up the low energy end of the X-ray spectrum has one or two drawbacks. The energy peaks crowd together and it is quite common for peaks to overlap. Examples are numerous but include those of the oxygen K (energy 523 eV) and chromium L lines (energy 573 eV). The energy resolution of the X-ray spectrum recorded with EDXS is about 80 eV at the energy of the example quoted and the O-K and Cr-L peaks are not resolved. A whole alternative technology exists that uses diffraction from crystals to disperse X-rays according to their wavelength, and its main attraction is the high energy resolution obtainable. The bulkiness of these detectors and the complexity of their mechanical construction has meant that very few are found in STEMs, although the earliest analytical instruments, such as EMMA-4, did use this technology (see Chapter 8 for some more details).

Energy resolution is a complicated question that we shall not pursue far in this book, but it depends in part upon the electronic circuits that detect the pulses coming from the diode (see Section 6.2.1). Measurement of the size of the pulse (and hence the energy of the X-ray that gave rise to the pulse) requires a certain amount of time (usually between 5 and 50 μsec); if the time allocated is short then the measurement is less accurate than if the time allocated is longer. Energy resolution is improved with longer 'process times' but at the expense of increased 'dead-time'.

6.5.3 Dead-time

As was described in the last paragraph, it takes a finite time to analyse each X-ray. During this time, another X-ray cannot be processed. The electronics contains circuitry to disable the detector during this period – we call it 'dead-time'. Unfortunately, even though the arrival of an X-ray during the dead-time might not interfere with the analysis of the first one, it does extend the dead-time. Hence, if the arrival rate of the X-rays is too high, the dead-time approaches 100%, and no X-rays are recorded. The maximum count rates achievable in the most modern analysis systems are of the order of a few tens of thousands of counts per second. Technology has been progressing fast enough, however, that, at the time of writing, many slightly older systems, capable of recording only a few thousand counts per second, are still in use. These count rate limitations are a significant problem in the SEM, where the specimen is thick and the probe is not necessarily very small. However, in the STEM, where a thin specimen and small probe are used, there will often be so few X-rays detected that the dead-time limitation will not be approached.

Dead-time can arise because of light leakage into the detector either from cathodoluminescence (see Section 6.2.2) or from within the microscope itself (e.g. light from a thermionic filament transmitted down the column and reflected into the detector). Excessive dead-time is also common when too many X-rays are generated by irradiating thick parts of the specimen or too high a probe current is employed. Dead-time does not cause damage to the detector.

6.6 Electron energy-loss spectroscopy

In the introduction to this chapter we mentioned briefly the principle of EELS; we will now discuss the technique more fully. EELS is potentially a rich source of information about the sample, including chemical binding and electronic structure, but is often difficult to interpret, and requires instrumentation that is expensive, and until quite recently was not commercially available.

As a high-energy electron passes through a specimen, a number of things may happen. One possibility is that it may emerge completely unaffected, but this is a rather uninteresting case. A second possibility is that the incident electron may be scattered elastically – in other words, its direction of travel (more technically, its 'wave vector') may change, but its energy stay the same. If the specimen is crystalline, we call this second process diffraction, but elastic scattering can occur in non-periodic structures too. The spatial intensity distribution of the scattered electrons (the diffraction pattern) depends principally upon the atomic structure of the specimen, and is a valuable tool for the microanalyst, as has already been discussed in Chapter 5.

A third possibility is that the electron can give some of its energy to the specimen, through one of various mechanisms. It is this possibility that is the subject of electron energy-loss spectroscopy. It happens, as we shall discuss, that careful examination of the energy spectrum of the transmitted electrons can yield a rich variety of information about the sample; before we get to that, though, we will describe the instrumentation typically used for EELS studies.

6.7 Energy loss spectrometers

By far the most common energy-analysing system in use on electron microscopes today is the magnetic sector. This depends for its operation on the fact that electrons, when travelling in a plane perpendicular to a uniform magnetic field, are constrained to follow circular paths whose radii depend upon the velocity of the electrons. *Figure 6.5* illustrates the principle of a simple spectrometer (see, for example, the black box labelled 'PEELS' on top of the VG STEM in *Figure 1.1b*). A beam cross-over in the electron column serves as the electron source for the spectrometer.

Considering for a moment only the electrons that have lost no energy, a divergent beam (whose angle is determined by the collector aperture, not shown in *Figure 6.5*) passes into the analyser. The geometry of the entrance and exit faces of the magnet is designed carefully to give a focusing effect, so that the electrons diverging in the plane of the drawing cross over again at the point shown. There is no focusing effect perpendicular to the plane of the drawing so the actual focus is a line rather than a point. Electrons of some lower energy might follow the tighter trajectory as shown, and are brought to a focus at a different place. Hence a suitably positioned slit will allow only electrons with a small

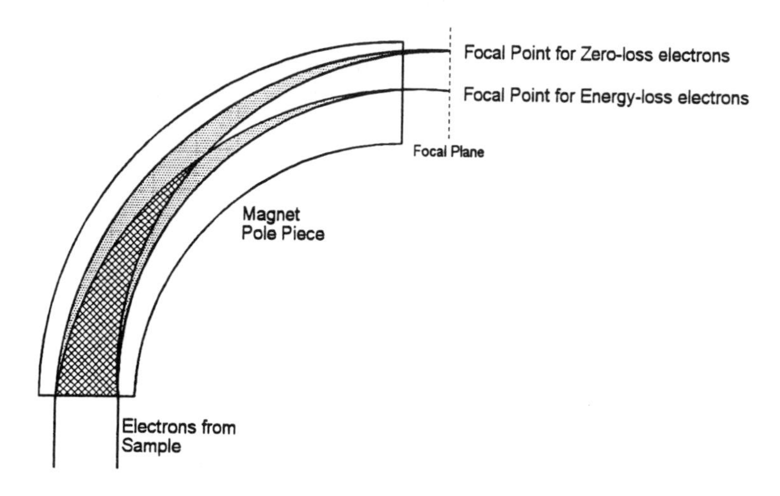

Figure 6.5. Illustration of the principle of electron energy-loss spectroscopy. The beam of electrons to be analysed is passed between the pole pieces of a magnet. The electrons travel in circular trajectories, the radius being a function of the electron energy. By careful design of the pole pieces (not illustrated here) the electrons can be brought to an energy-dispersed focus in a plane, in which can be placed a slit or spatially resolved detector.

energy range to pass (the exact energy can be adjusted by varying the strength of the magnet, and the range by adjusting the width of the slit).

This type of spectrometer can be incorporated (in more sophisticated form) into an imaging microscope, either between the objective and projector lenses or as an attachment mounted at the end of the column, resulting in an 'energy-filtered electron microscope', in which the image viewed on the screen is formed by the energy filtered electrons. While this type of instrument can perform many of the same type of experiments as those performed in scanned-probe instruments, and, indeed, can do some things either better or more conveniently, it is not a STEM, and so, though an important type of system, is outside the scope of this book.

Modern practical 'bolt-on' spectrometers also include a number of lenses for focusing and magnifying purposes. Most are capable of an ultimate energy resolution of about 0.4 eV, which is smaller than the energy spread of the electrons coming from a thermionic electron gun, and close to the energy spread of a field-emission source. It is also close to, or better than, the specification for the stability of the high-voltage supply on many intermediate-voltage microscopes – it is a tribute to the microscope manufacturers that spectra can be recorded to this precision!

6.7.1 *Interfacing to the microscope*

An electron which loses energy in a collision with an atom will in general also have its direction changed slightly; the greater the energy loss, the larger the average deflection angle. While the probability (known as the 'cross-section') of scattering at a given angle is readily computed, the difficulty in practice is in knowing which electrons enter the spectrometer and which do not. The entrance to the spectrometer is generally defined by a moveable aperture, whose size is known. The difficulty arises because the specimen is deeply immersed in the magnetic field of the objective lens. The portion of the field after the specimen causes the divergence angles of the transmitted electrons to be compressed (see Section 5.2.4). Changes in height across a fully tilted specimen can lead to quite large changes in post-specimen compression (often by as much as a factor of two), leading to major uncertainties for quantitative EELS analysis.

The problem is especially acute in VG HB501 and HB601 microscopes, because the vast majority of these instruments do not allow the possibility of adjusting the specimen height in the lens. Most TEM/STEMs and the VG HB603s do have specimen height adjustment, but they all share another problem – these instruments all have a series of projector lenses following the specimen, and these lenses must be excited to form a cross-over in the beam at the focal position of the spectrometer. Unfortunately, all lenses suffer from chromatic aberration – the inability to focus electrons of different energies at the same point. Hence, while the spectrometer may be focused perfectly for the zero-loss electrons, it will be progressively more out of focus for the higher energy losses. These difficulties can be overcome, but only by careful calibration of the microscope and spectrometer, and by painstaking attention to detail during the analysis.

6.7.2 Data collection systems

With the design of spectrometer just discussed, an electron energy spectrum can be obtained by changing the excitation of the spectrometer magnet and recording the intensity of the electrons transmitted through the slit. This type of detector system is slow because it is a serial system, though it was the only one available for many years and much important work was done with it. However, it was realized that large amounts of information were being lost on the slit. For example, if the data were being recorded in 1024 channels, then each particular energy loss intensity was being recorded for only 0.1% of the time of the total experiment, or, to put it another way, 99.9% of the signal was lost!

The need to improve the detection efficiency led to the development of practical two-dimensional detectors which allow the whole spectrum to be recorded simultaneously, or 'in parallel'. These systems became commercially available from the mid-1980s (soon after the advent of the windowless EDXS), and are termed 'parallel electron energy-loss spectrometers', or PEELS systems. The earlier, serial type of spectrometer, formerly known simply as an 'EELS' system, is now usually given the acronym 'SEELS'. Apart from the problem of loss of information, the SEELS system had other difficulties, mostly related to the extreme dynamic range of the energy loss spectrum (the electron intensities at different parts of the spectrum can differ by a factor of 10^6 or more). These difficulties also exist in PEELS systems, although in different forms. The advantages of the PEELS greatly outweigh any disadvantages, so they are the system of choice for the great majority of investigators.

It is perhaps important to note that although today virtually nobody buys a SEELS system, there are functions, especially on a STEM, for which it was superior (or which in some cases a simple PEELS system cannot perform). The formation of energy-filtered STEM images or diffraction patterns is an example, since the slit mechanically selected an energy window. As the manufacturers develop their products, this functionality will undoubtedly be added to PEELS, with some combination of hardware and/or software.

6.8 Energy transitions

Let us now return to discuss the interactions of the electrons with the specimen. It is important to remember that bound systems generally have quantized energy states – in other words, only discrete energies are possible. Hence the energy required to excite the system from one state to the next is well defined – the exciting incident electron can give neither more nor less energy to the specimen than the difference between the energies of the two bound states. Alternatively, if the system is excited from a bound state to an unbound state (an example would be the ionization of an atom by the ejection of an electron) then the incident electron must give up at least the binding energy, plus the kinetic energy of the ejected electron.

In conductors, there are free electrons which can be given any amount of energy. In general these excitations give very little information about the sample, and are regarded as a nuisance. Another low-energy excitation is called a phonon, which is a lattice vibration of the atomic nuclei – essentially a sound wave in the specimen. Although quantized, the phonon energy is below the resolution of the spectrometer, so cannot be used to obtain information about the specimen.

Excitations of electrons across the band gap in semiconductors and insulators (which are, of course, a form of ionizing event) occur in the energy range of a few electron volts, and can be resolved by the spectrometer. While the fine details of the structure of these events is often obscured by the limited resolution of the spectrometer system (and the tails of the zero-loss peak), careful study can reveal important differences in the electron states at, for example, grain boundaries, compared with the bulk material. Such experiments are in their infancy, but show great promise for the future.

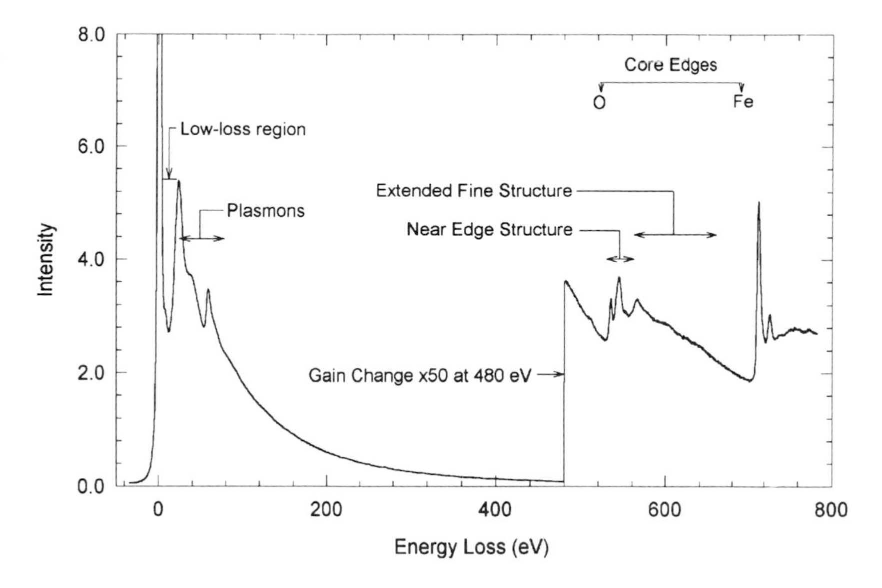

Figure 6.6. PEELS spectrum from a thin section of an iron meteorite. The various regions of the spectrum are indicated (but note that the term 'low-loss region' is often applied to the entire spectrum below 100 eV loss). The spectrum was recorded at 400 kV, and is reproduced by courtesy of David C. Bell, Massachusetts Institute of Technology.

In the energy loss range around 10–15 eV are observed plasmon peaks. Plasmons are oscillations of the entire electronic structure of the solid, and are very strongly excited, to the extent that they are usually the strongest features observed in an EELS spectrum, after the zero-loss electrons (the electrons that pass through the specimen without interacting with it). *Figure 6.6*, which is a spectrum recorded from a ferrous meteorite, illustrates plasmon peaks, and is described more fully in the next section.

Frequently an electron might excite two or three plasmons as it passes through a specimen, so the energy loss spectrum will show this number of peaks at multiples of the plasmon energy loss. The plasmon energy can be quite sensitive to the composition of the sample, so can in principle be used as an analytical tool; however, since we only have a simple analytical method for predicting the plasmon energy (based on free-electron density) – a calibration must be carried out with known samples of composition near that of the unknown material. While there are specific cases in which measurement of the plasmon energy has been successfully used for analysis, it is not generally a practical method.

The majority of EELS investigations make use of the so-called core-loss events which occur when an incident electron ionizes one of the more tightly bound electrons of an atom in the specimen. To a first approximation at least, the probability of this ionization occurring is independent of the environment of the atom; hence, the signal, in principle, can be used to provide quantitative analytical information. Before going on to discuss how this can be achieved, and some of the difficulties of EELS analysis, however, we will describe how the various energy-loss events we have described appear in a spectrum.

6.9 Details of the energy loss spectrum

If we imagine a specimen so thin that almost all the electrons pass through – in other words, the probability that an individual electron will lose energy is small, then we can see that the probability of an electron undergoing two energy-loss events can be neglected (since it is the square of the probability of a single event). In this case, if we plot the number of electrons detected at different energy losses, the result simply corresponds to the probability of an electron losing that energy (and, we should add, passing through the spectrometer, but more of that later). Referring again to *Figure 6.6*, we observe that the dominant feature is the zero-loss peak, indicating that a substantial number of electrons are not scattered inelastically, as our description requires. We have indicated on the figure the low-loss region in which the band-gap events would be seen, although they are not illustrated here. Next, at a somewhat higher energy loss are seen the plasmons. (Our terminology is probably a little confusing here – many would use the term 'low-loss region' to include all energy losses up to 100 eV, despite the fact that 'core losses' can be observed at smaller energy losses.) The intensity of the electron signal at larger energy losses becomes very small, so the remainder of the spectrum is shown at 50 times the scale. At this scale the signals due to the major elements O and Fe are visible as steps in the background which otherwise decays with increasing electron energy loss. The energy loss at which the step occurs is the ionization energy of the particular atomic electron involved in the interaction.

The transmitted electron can lose any amount of energy above this minimum; this is illustrated in *Figure 6.7*, which was recorded from a boron nitride flake (which seems to have oxidized!) supported on a carbon film. The background before the boron edge is extrapolated to higher energy losses; it can be seen that the actual signal always remains above the extrapolated background. The same would be true if the background below, for example, the carbon edge were modelled and extrapolated. The background before the first of the core-loss edges arises, amongst other possibilities (e.g. bremsstrahlung), from the valence electron ionizations and, in the case of a conductor, excitations of the conduction electrons.

It is apparent from *Figures 6.6* and *6.7*, and is generally true, that the probability of an energy-loss event occurring is roughly inversely proportional to some power of the energy loss. Hence above about 1 keV loss the signal becomes very small, and there are few times when EELS at energy losses above 2 keV are attempted. However, all elements have

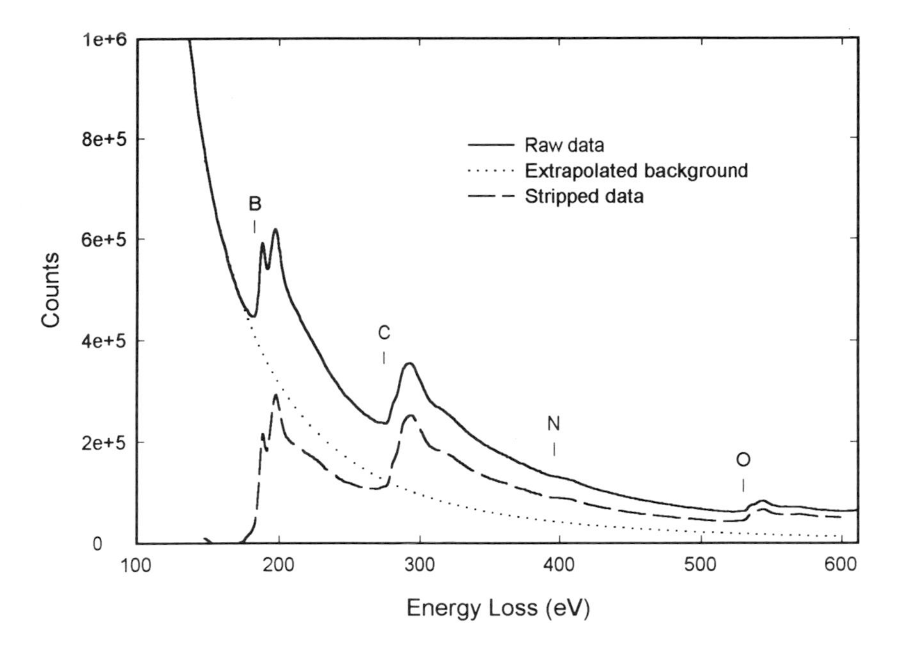

Figure 6.7. PEELS spectrum (recorded at 250 kV) from a boron nitride flake supported on a thin carbon film. The background before the boron edge has been extrapolated and stripped from the spectrum, illustrating that each edge contributes signal to the whole energy-loss range above it. Interestingly, this spectrum appears to indicate that the sample, which is over twenty years old, has oxidized in storage.

states that can give rise to energy-loss events in this range, so are detectable (though, as a practical matter, hydrogen in solids and lithium in alloys are very difficult to interpret).

In our discussion so far we have described the excitation of electrons from bound states as though these states have only a single, well defined, and unalterable binding energy. In fact, in the majority of cases this description is not accurate.

6.9.1 Near-edge structure

It is beyond the scope of this book to describe the physics involved, but by careful study of the details of the shape (or 'structure') of the edge in the energy loss spectrum, it is possible to derive significant information about the binding of the atoms in the solid. A simple example is shown in *Figure 6.8*, where we compare the K edge of carbon recorded respectively from diamond and from an amorphous carbon film. In this case, inspection would allow us to determine whether a third spectrum from an unknown were closer to the diamond or to the amorphous material.

For completeness, we must mention here that carbon may also show graphitic bonding, which gives a different characteristic shape to the EELS edge. In more complex cases, careful band-structure calculations can allow a solid-state physicist to determine quite precise details of the

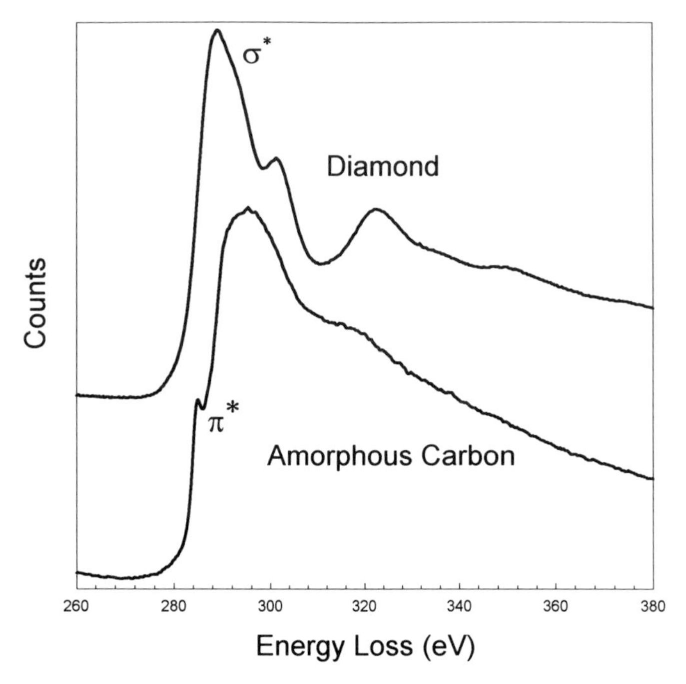

Figure 6.8. PEELS spectra of diamond and amorphous carbon, recorded at 400 kV, overlaid to illustrate the difference in the shape of the edge. Comparison of an unknown spectrum with such calibration spectra allows, in many cases, the chemical nature of the elements in the sample to be deduced. Reproduced by courtesy of David C. Bell, Massachusetts Institute of Technology.

chemical bonding from the details of the EELS spectrum. For example, the structure of the iron edge in *Figure 6.6* could be compared, either with theory or with spectra recorded from well-known samples, in an attempt to determine the stoichiometry of the material. For obvious reasons, this type of study is known as 'energy-loss near-edge structure', and is given the acronym ELNES.

6.9.2 Extended fine structure

Still more information can be derived by examining the structure of an edge for several hundred electron volts beyond its onset. This type of structure is called 'extended energy-loss fine structure', and is given the acronym EXELFS. The structure arises from, and therefore leads to information about, the local atomic environment of the target atoms, but the information is not easy to obtain. A spectrum with extremely high signal-to-noise ratio is a prerequisite, and while relatively uncomplicated Fourier analysis is employed in the data manipulation, the interpretation depends strongly on knowledge of details of the sample. *Figure 6.6* also shows extended fine structure on the oxygen and iron edges, although one would hope for better signal-to-noise if one were going to make a careful study of these edges.

6.9.3 Specimen thickness effects

The perspicacious reader will have noticed our emphasis on 'thin' specimens, without our having put figures on the description. When is a specimen thin enough? In the preceding section, we assumed that the 'majority' of the electrons pass through the specimen unaffected. This is clearly an extreme case. Consider if the specimen thickness is equal to one mean-free-path length for inelastic collisions, then about 40% of the electrons pass through the specimen undeflected. Unfortunately, by the same criterion, only 40% of the electrons that suffer one energy-loss event pass out of the specimen without suffering at least one more! We can see from this simple argument that one mean free path is about the upper limit of useful specimen thickness. The mean free path varies with the sample material and the electron beam energy, but for 100 kV electrons, is usually in the range of a few tens of nanometres (which is 'quite thin' for a typical TEM specimen). The mean free path increases roughly in proportion to the electron energy, so the specimen can be thicker in an intermediate-voltage (200–400 kV) microscope, but the higher energy can bring other problems, such as greatly increased beam damage. If the specimen is suffering any contamination (build-up of carbonaceous material under the beam) the thickness can increase rapidly to the point where EELS is impossible. In such a case, EDXS analysis might still be quite satisfactory, except that the spatial resolution would be degraded. The effects of the multiple scattering, resulting from too large a specimen thickness, can be partially mitigated by numerical processing using Fourier techniques, but even so a practical upper limit on specimen thickness is not much greater than one mean-free-path length.

6.10 Detection limits

We have not mentioned the elemental detection limits for PEELS (SEELS limits are much poorer). While qualifying our statements by adding that these are ideal cases, there are reports appearing to show the detection of individual atoms. One can hardly expect a smaller minimum detectable mass! Turning to minimum detectable mass fraction, PEELS has been shown to be capable of detecting well below 1 wt% of a wide variety of elements in standard glasses, albeit in a microscope fitted with a LaB_6 thermionic electron gun, which is capable of generating a larger current (but also in a larger probe diameter) than can the field emission gun. There have also been calculations that suggest that PEELS in such a microscope is more sensitive than EDXS for all elements lighter than calcium.

6.11 Analytical strategy

In this section we will briefly discuss how to select an analytical strategy for a sample.

In general, EDXS analysis is simpler to use for qualitative and semiquantitative elemental analysis. It is easier for a novice user to set up, and, while not free from artefacts and pitfalls, can provide useful information in a relatively short time.

PEELS analysis requires a carefully selected thin area of the specimen, and careful set-up of the microscope. It may be an advantage to use PEELS analysis to resolve peak overlaps (see Section 6.5.2), but the main advantage of PEELS is in its sensitivity to chemical bonding.

In this chapter we have concentrated on point analysis – the beam is placed on a point of the specimen and the spectrum recorded. This technique is appropriate when the areas of interest are readily identifiable, but is very labour intensive when the general positional relationships between the different elements are being investigated. An alternative analytical strategy involves the use of computer control to derive a two-dimensional plot, or 'map', of the elemental distributions. This method is the subject of the next chapter, and is useful either with EDXS or PEELS analysis.

6.12 Conclusion

EELS, unlike EDXS, is not a technique that can be used successfully by a novice. The experienced user with good insight into the problem can derive much chemical information by comparison of an unknown spectrum with carefully selected standards, while a theoretical physicist can add still more to the range of available knowledge by comparing theory with experiment. The results go in proportion to the effort expended!

7 Mapping in the STEM

7.1 Introduction

By 'mapping' in the STEM we mean displaying in two dimensions the distribution of (usually) an element or elements in a region of a sample. (We could make a case that the bright and dark-field images are 'maps' of the respective signal intensities, but more of that later.) Any signal conveying information can, in principle, be mapped, but we will limit our discussion to EDXS, EELS, and high-angle annular dark-field signals. Maps (especially those acquired with modern instrumentation) allow the operator to locate areas of interest for more detailed quantitative analysis, and they frequently provide a powerful and persuasive illustration of elemental distributions which, in a presentation, can make a scientific point to a sceptical audience more easily than many spectra and paragraphs of explanation. Thus, although mapping is, as we shall see, frequently not quantitative, it is a very commonly used data acquisition technique. As in the last chapter, we will begin by discussing the technique which has been in use longest, EDXS mapping.

7.2 X-ray mapping (including linescans)

The earliest systems for X-ray mapping produced what we now refer to as 'dot ' maps. In this method, what we call a 'window' is defined in the X-ray energy range; using either hardware or software methods, a pulse is passed whenever an X-ray is detected whose energy lies within the range defined by the window. If this energy range corresponds to the characteristic peak of an element, then the pulses will be detected more frequently when the probe is on areas of the specimen containing that element. As the electron probe is scanned slowly across the specimen, the spot on the imaging cathode ray tube (CRT) is scanned synchronously (just as when imaging in STEM). The CRT beam is 'blanked' (turned off) except when a pulse is received from the X-ray analyser, so whenever an X-ray is detected a spot of light is generated on the screen. A photographic film is exposed for the duration of the scan, recording the positions of all the dots, leading to the name given to the map. If more than one element is of interest, then a second exposure is made, with the window redefined to the energy of the X-rays of the second element, and so on. As tedious as this sounds it was the only way to proceed with analogue systems.

Clearly, the area-density of the spots will be higher in areas containing the element of interest. There are a number of drawbacks to this system, but until the advent of inexpensive computers (and cheap memory) it was widely used. Rather than enumerate the difficulties, we will, in the next section, mention the improvements brought about by the advent of full digital mapping.

7.2.1 Digital mapping

Digital mapping, as implemented on the majority of STEMs today, relies on control of the probe position on the specimen by the computer which is also processing the X-rays. A number of windows corresponding to the X-rays of interest are defined and each window's map is stored separately. The digital method provides faster response but there is a limit to the number of possible windows. When defined in software the number of maps can be large. However, with modern fast computers, the speed advantage of the hardware systems is less significant now than it used to be a few years ago. The probe is scanned over an array of discrete points ('pixels'), pausing for a predefined time (dwell-time) at each point. The number of counts recorded in each window during this time is stored in memory (and may simultaneously be displayed in segments of the monitor).

Since the computer has access to the 'live-time' information from the X-ray analyser, it can correct the pixel dwell-time, so that if the specimen thickness varies substantially then the real dwell-time can be varied. Clearly, the maps for all the elements are acquired simultaneously, unlike dot-mapping where the specimen must be scanned for each element. The intensity resolution at each pixel is determined by the number of bytes of memory allocated to storing the number of counts. Typically this would be a single byte, allowing up to 255 counts (8 bits) to be recorded, but larger numbers could be stored where appropriate. The system is not limited to recording X-ray information. Any other available signal, such as the bright or annular dark-field STEM image can be recorded simultaneously. Most commercial X-ray analysers have made provision for digitization of at least one electron image input.

Since all the maps (and images) are acquired during a single scan, they are of course in perfect register, even if the image is not absolutely stable. This is an issue especially in a STEM where maps may be acquired at very high electron-optical magnifications, where some small amount of specimen drift is inevitable, and if a thin sample is being examined, or trace elements sought, extended acquisition times may be used. A number of efforts have been made to have the controlling computer examine the position periodically and correct for drift. While these programs certainly do have utility, they are not a panacea for all situations. In particular, in the highest magnification STEM images, there is often insufficient sharp detail for them to be effective, and even very modest amounts of contamination build-up can confuse them utterly.

The resulting arrays of X-ray intensity data are, in principle, exactly like images, and it is very convenient to store them to disk in a common image format. They can then be read into a common commercial image presentation or analysis program for processing for display or to derive quantitative information.

Figure 7.1 shows a case where the maps present in visual (and readily interpretable) form the relationships between the elements present. The sample is of magnesium powder, and the purpose of the investiga-

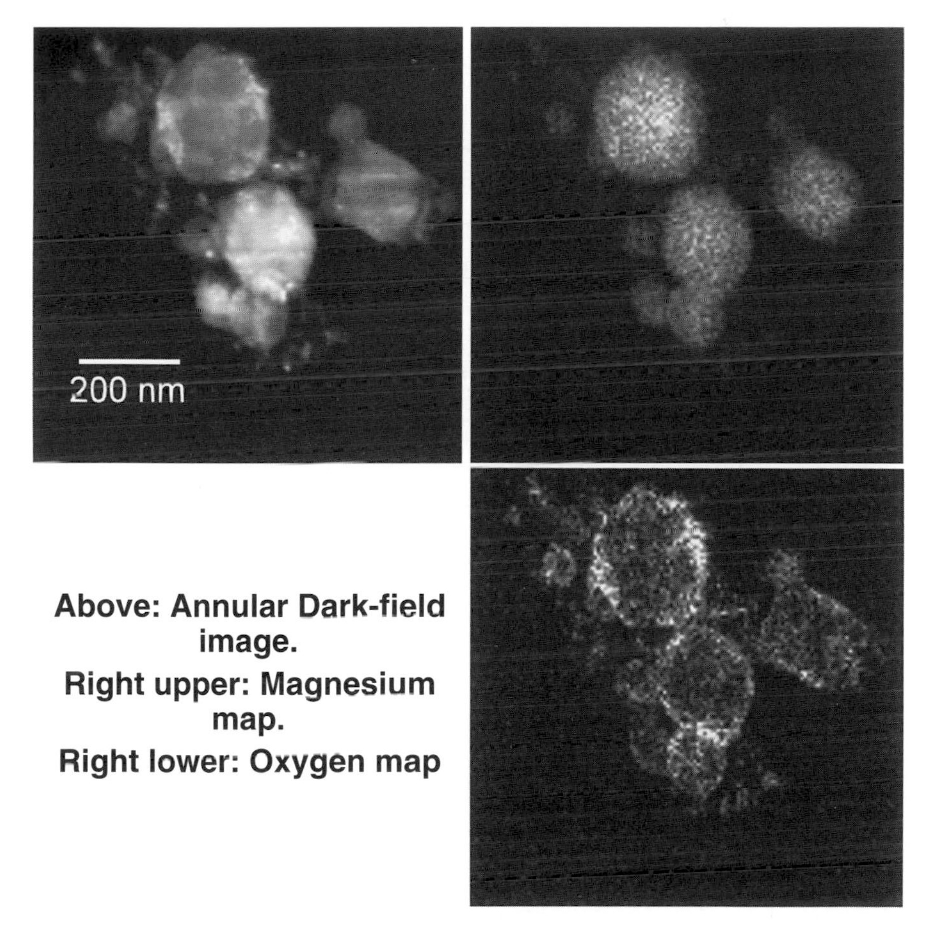

Figure 7.1. Annular dark-field image, and oxygen and magnesium maps of processed magnesium particles. Comparison of the oxygen and magnesium distributions allows one to deduce that the oxygen is present only on the outer surface of the larger particles. (Colour overlays would show this dramatically.) Reproduced by courtesy of Alex Diaz and Adele Sarofim, Massachusetts Institute of Technology.

tion was to evaluate different material processing parameters. In this case there was reason to suppose that, with the Mg, there was also present C, O, and Si. The powder was dispersed on a carbon support film on a copper grid, and analysed. An integrated X-ray analysis of a small cluster of particles confirmed the elements present (except that carbon was known to be present in the support film). The X-ray map in *Figure 7.1* clearly shows that the larger, well-defined particles, are of magnesium metal, with some slight oxygen; the smaller, less well-formed particles are principally MgO (there are also silica and magnesium carbide phases present, but we have not illustrated these). Careful study of the relative intensities of the oxygen and magnesium signals across the Mg particles leads to the conclusion that the oxygen is present in a thin oxide coating on the metal. It would have been virtually impossible to derive this information from a number of point analyses. This conclusion is very clear in a colour combination of the elemental maps, but this cannot be illustrated in this book.

Other examples of applications of X-ray mapping have included derivation of the volume and number density of ZrO particles on the surface of an alumina scale growing on a high temperature alloy; determination of the size and number of Si precipitates in fine Al–Si wires; demonstration of the location of metal-containing phases in metal-loaded polymers; and illustration of the different layers formed in a complex corrosion scale formed on a stainless steel.

7.2.2 Digital linescans

A special case of a map is a linescan, where the probe is scanned, not over an area of the specimen, but along a single line, usually across some feature of interest. In a map, the precision of the analysis is very limited, because of the short dwell-time at each pixel. For example, a map acquired at a pixel resolution of 128 × 128 contains 16 384 pixels; if the dwell-time on each pixel is 50 msec, then, allowing for some analyser dead-time (and disk storage time), the acquisition time is approaching 20 min. Considering that a typical quantitative point analysis might be acquired for 100 sec, it can be seen that the amount of statistical data acquired per pixel must be very low. A 128 point linescan, though, acquired for the same time, would allow the pixel dwell-time to be about 10 sec, giving far more data per pixel.

An example of a study employing linescans was given in *Figure 1.2*. Of course, poor definition of detail in a linescan could be caused by the probe being out of focus, by beam broadening in an excessively thick specimen, or by poor specimen preparation (amongst other things). It is therefore essential that the analyst be alert to these difficulties and ensure that the microscope is set up properly, if critical deductions are to be made from such results. Unfortunately, even the best microscope has some instability, and even in a linescan the time taken to acquire the data can become prolonged if high-precision data are to be obtained.

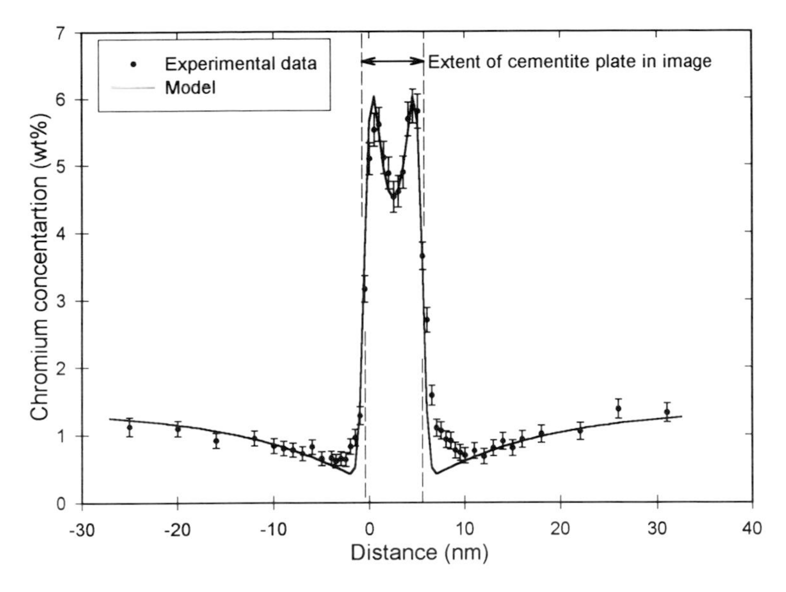

Figure 7.2. Manually recorded linescan plotting chromium composition as a function of position across a cementite plate in chromium pearlite. The data are compared with a model derived by convoluting a Gaussian probe shape with a full width at half maximum of 1.2 nm with a plausible model of the actual distribution, adjusting parameters for best fit. It is evident that there are 'tails' on the probe extending about 2nm beyond the Gaussian centre. The microscope was a VG HB603 operating at 250 kV.

Hence, some form of manual intervention becomes inevitable. The extreme is illustrated in *Figure 7.2*, which shows a plot of the chromium composition measured (at 250 kV) across a cementite plate in chromium pearlite, superimposed on a model derived from a convolution of a Gaussian beam distribution of full width at half maximum of 1.2 nm with a plausible model of the actual chromium distribution derived from diffusion theory, with parameters adjusted to give a good fit to the data.

These data were acquired entirely by hand. Many attempts were made, without success, to duplicate this result under computer control – a reflection, perhaps, that even today there is no substitute for a skilled operator! We note that the relatively poor fit in the illustration between the data and the model just outside the cementite plate probably arises from the presence of spherical aberration 'wings' on the electron probe.

7.3 High-angle annular dark-field imaging (HAADFI)

We recall from Chapter 4 that the annular dark-field image is formed by the electrons that are scattered in the specimen. It happens that different scattering processes tend to scatter the electrons into different angular ranges; therefore, by selecting the angular range detected, we can form images that give us information about the particular process of interest. The selection of the angular range is comparatively simple: an annular detector of appropriate dimensions is used. In a microscope with lenses after the specimen, there is added flexibility, because these lenses can be adjusted to change the collection angles of the detector (see *Figure 4.4*).

A typical HAADF detector may collect electrons scattered only over the range from about 2.8° to 8.6° (50–150 mrad), which might seem like small angles, but the technique is given the name 'high-angle scattering' because most other scattering processes lead to scattering through smaller (often much smaller) angles.

The main process scattering electrons through higher angles (> 50 mrad, or > 2.8°) is scattering by the nuclei of the atoms in the specimen. The probability of an electron being scattered through such high angles by an atom is approximately proportional to Z^2 (Z is the atomic number of the atom). The total scattered intensity at a given angle depends, of course, on the number of atoms, which in turn depends on the specimen thickness and density, but in a region of sensibly constant thickness the signal is strongly dependent on the average atomic number of the atoms in the sample.

HAADF images do not uniquely identify the atoms giving rise to the image. It has been estimated that the 'Z' resolution of HAADF imaging is about 20%. Thus if two regions of the specimen have compositions that happen to have similar average values of Z, then they will not give rise to significant contrast in the image (e.g. GaAs and InP have average values of $Z = 32$). Hence the method cannot be used alone on unknown samples. However, there are many cases where a HAADF image can provide a qualitative image of elemental distributions very quickly.

The applications described above are interesting, but HAADF imaging as described so far is clearly of rather limited utility. Consider what happens if the electron probe is made very small. In VG601 and 603 STEM instruments (and reportedly in other companies' instruments under development), the probe can be reduced to the 0.15–0.2 nm range. The current in this probe would be so small that X-ray analysis would not be possible and even PEELS would be hard pressed at this resolution. It has been shown that atomic columns in crystals can be imaged by HAADF (see *Figure 4.8b*), and that the resulting images can be readily interpreted (unlike images obtained by bright-field microscopy in STEM or TEM). Furthermore, the intensity difference between different columns can indicate composition differences on an atomic scale. Not only that, but defects (such as grain boundaries) can be imaged, and the local rearrangement of the various species may be deduced.

High-resolution HAADF imaging is quite new, and is still a method under active development, most notably in the United States by groups working at Cornell University, IBM, and Oak Ridge National Laboratory. As can be seen from the foregoing description, it promises to be extremely powerful, especially when combined with image analysis techniques.

8 Limits to STEM and advanced STEM

Over the past 30 years, image resolution in both CTEM and STEM have asymptotically approached a value of between 0.1 and 0.2 nm (depending on the resolution criterion adopted: point resolution, line resolution, or information limit). This reflects the fact that we are approaching a limit set by physics and the very practical necessity of being able to take a specimen in and out of the microscope.

8.1 Limits to microprobe analysis

In *Figure 8.1* the minimum number of atoms that could be detected using bulk or thin foil X-ray microprobe analysis is plotted as a function of year. A big step in spatial resolution for microprobe analysis accompanied the transition from the use of bulk to thin specimens between 1960 and 1970. The first transmission electron microscopes in which it was possible to carry out microprobe analysis, the EMMA series (electron microscope microprobe analyser), were developed in the 1960s. In the original EMMA instruments X-rays were analysed with wavelength

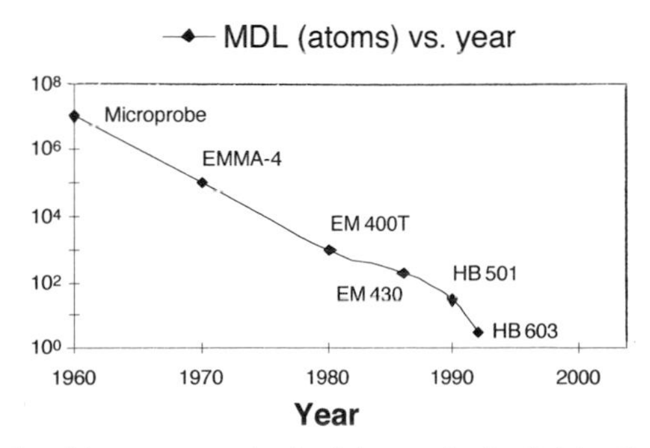

Figure 8.1. The rate of improvement of minimum limit of detection (MDL) during the years 1960 to 1993.

dispersive X-ray spectrometers (WDXS) and, in order to provide egress of the X-rays, a long focal length Le Poole mini-lens, with a spherical aberration coefficient (C_s) of 4.2 cm, was used to focus the electron probe. As a result of the large C_s value (and the tungsten thermionic source), the minimum probe diameter, which contained a current from 10 to 100 nA, was about 130 nm.

Since 1970 the spatial resolution for analysis in the AEM or STEM has changed dramatically. The high X-ray collection efficiency of EDXS (0.05–0.3 sr compared to 0.003–0.01 sr for WDXS) short focal length probe-forming lenses with C_s values of a few millimetres and high brightness FEG sources has enabled a current of up to 1 nA to be focused into analysis probes less than 1 nm in diameter. Thus for example the current density in the VG601UX STEM probe is over 150 times that available in its ancestor EMMA-4.

It is of interest to note that the extrapolation of the data in *Figure 8.1* to one atom occurred in the mid-1990s: virtually single atom detection using EDXS detectors in the FEG-STEM is with us now.

The data in *Figure 8.1* are for the detection of X-rays. In principle EELS is a more efficient technique than EDXS. Even with the 'large' solid angle of collection which can be obtained with the EDXS detector positioned very close to the specimen, a solid angle of 0.2 sr means that the detector only intercepts about 1.6% of the characteristic X-rays produced in the specimen. In EELS the majority of the electrons which have suffered a characteristic loss in energy enter the detector. Unfortunately the peak-to-background ratio in EELS is inferior to that in EDXS because of the higher background signal. Nevertheless, parallel EELS signal collection, PEELS, employing sophisticated signal processing has shown, two or three years earlier than with EDXS, that single atoms can be detected.

Under ideal conditions, using a FEG source and with careful processing of the spectra, both EDXS and PEELS are capable of detecting single atoms. Future improvements in the FEG instrumentation, operating at higher voltages (200–400 kV) and more sophisticated data processing software will **not** mean that it will be possible to detect half an atom! However, the sensitivity and ease of detection of trace elements will continue to improve.

8.2 Developments of the FEG

As pointed out in Chapter 4, to a first approximation the resolution in STEM is determined by the probe size. As the probe at the specimen is a demagnified image of the source, the smaller (and brighter) the source the smaller the probe and the finer the detail that can be resolved in the image. A field emission source, which is used in many modern AEMs, is the key to the formation of the small probe which contains a sufficient number of electrons to produce a high quality image or a high resolution microanalysis. It is unlikely that completely new sources will be developed to replace the FEG, but it is reasonable to expect that incremental improvements will continue to be made to the FEG.

For example, one (S)TEM manufacturer has produced a 'Schottky' source that is reported to be able to produce a maximum probe current between 100 and 200 nA (about 20 times that obtainable with a cold FEG). Furthermore there are reports that coating tungsten with zirconia (to reduce the work function) increases the brightness and sharpens the energy spread of cold field emission sources.

The amount of current drawn from a thermionic source is directly proportional to the accelerating voltage: if the accelerating voltage is doubled this leads to a doubling of the current drawn from the source. The relationship is more complicated for a FEG source, but a significant increase in electron current density accompanies an increase in accelerating voltage. There are several other advantages of increasing the accelerating voltage which include increased specimen penetration, higher peak-to-background ratios in X-ray spectra, shorter electron wavelengths, and an increase in the mean free path for electron energy losses in the specimen, which leads to improved EELS analysis. These advantages have been recognized in the development of the current generation of higher voltage (200–400 kV) (S)TEM instruments. Unfortunately the cost of the instruments increases not linearly with voltage but at a significantly higher power, as do the practical problems associated with engineering of high voltage and lens supplies.

8.3 Spherical aberration

As discussed in Chapter 4, one of the fundamental limits to probe size is the spherical aberration coefficient of the probe-forming lens, C_s. The probe size is roughly proportional to C_s. One of the 'holy grails' of electron microscopists has been to design an electromagnetic lens in which spherical aberration is fully corrected. The same problem has been effectively solved by the optical microscopists by shaping ('figuring') the surfaces of the lens. Designs have been produced for a spherical aberration-corrected electromagnetic lens, but it involves the use of several other lenses, some of very complicated design. It would be very difficult to incorporate these 'corrected' lenses into a commercial (S)TEM and for the foreseeable future we will have to cope with the limitations imposed by spherical aberration.

8.4 Operation of the STEM

The VG STEM is a superbly flexible imaging and analysis tool, but it cannot be considered to be an 'easy' instrument to operate. As VG no longer manufactures microscopes anyone wishing to purchase an instrument on which STEM can be carried out is advised to purchase it from a company with a history of making TEMs and (S)TEMs. The conventional instrument manufacturers who exist today have survived, in part, because they have developed 'user friendly' instruments which are moderately difficult for the over-enthusiastic student to seriously damage. The operation of these modern instruments is managed by sophisticated software packages, which can be of great assistance to the operator in correctly setting up the instrument (including the alignment of the TEM and (S)TEM modes) to extract specific information from the specimen. An integrated approach to microscopy and analysis which draws heavily on computer control is the way forward, and the key to success is to be found in dedication to software development.

In the future the use of an instrument should permit data to be collected, analysed, stored, printed, and transferred to remote locations for 'off-line' processing. Full computer control in specimen manipulation is a new concept that promises much in terms of making the job of analysis more or less standardized. The hardware exists to provide the data that feed the software models that researchers use to understand the problems they are facing. What is required urgently is a means to gather the strands of EDXS, PEELS, and high-resolution EM together in a way that makes sense to our reader. Further reading on this subject is strongly recommended.

8.5 Spectrum imaging

A revolution in data handling has accompanied the increase in the computing power that is available at, or close to, the microscope. There will continue to be an increase in computing power in the immediate future and this will be reflected in the sophistication of data presentation. Most microscope users are familiar with X-ray mapping with SEMs, TEMs, or electron microprobes (see Chapter 7). Simple manipulations of the EDXS data collected at each data point are required: to subtract the background, apply k factors, and to display these data as weight fraction maps.

The spectrum imaging concept is an extremely powerful one, where whole spectra are collected at each data point in a map, and advanced spectral processing software has been developed to produce chemical maps 'off-line'. In an X-ray spectrum-image, for example, one can go back to the data and search for the presence (or distribution) of an element which was not thought to be of interest at the time of the analysis. The computer memory requirements can be prodigious – a 64 × 64 pixel X-ray image would require over 21 megabytes, since each spectrum occupies about 5 kilobytes. If each spectrum was acquired and saved to disk in 200 msec then a single spectrum-image would require over 3000 seconds (almost an hour). Recalling and analysing the spectra would take a similar amount of time.

It is possible to acquire the spectra to RAM, and save them to disk as a single file (a much faster process). However, a 512 × 512 pixel spectrum-image, storing 4 bytes of information from each of 1024 channels (such as might be used in a PEELS spectrum), would require a gigabyte of RAM and an equal amount of disk space – even with today's computers, this is not yet a practical requirement for a routine technique. Furthermore EELS spectra may require deconvolution to account for effects of multiple scattering to extract the data required and X-ray spectra may need corrections to be applied for different thicknesses.

Despite these complications PEELS spectral imaging has enormous potential because of the relatively short acquisition times compared to EDXS. The information that is contained within the PEELS spectrum-image can be extracted, manipulated, and mapped. This includes chemical composition from the energy loss edges, bonding information from the energy-loss near-edge structure (ELNES), and specimen thickness at each pixel-point. One ought to collect both EDXS and PEELS spectrum-images to be really thorough.

As has been implied, spectral imaging is not yet a routine technique, although spectral 'linescans' have been applied to the study of grain boundaries. With the continuing falling price of computing power and memory, it seems fair to conclude that in a few years time, the analyst will have this added technique in the available arsenal of tools.

8.6 Beam damage and drilling holes

The incident electron beam can cause radiation damage to the specimen either by the displacement of atoms or by radiolysis, the transfer of energy from the electron beam to the specimen by inelastic scattering processes. In many ceramics and minerals the damage rate under even moderate electron fluxes can be so rapid that extrapolation techniques (to zero irradiation time) must be used in order to carry out a quantitative analysis. During the analysis of silicate minerals such as feldspars (framework aluminosilicates of Na and Ca), Na loss due to radiolysis is well known during bulk microprobe analysis using a thermionic tungsten source. In a FEG-STEM the current density is several orders of magnitude larger than in a microprobe and damage occurs extremely rapidly (see *Figures 8.2* and *8.3*). Most silicates can be reduced to SiO_2 if they are exposed to a focused probe in the FEG-STEM for an extended period of time and many metal oxides are likewise reduced to the metal.

In metals electron beam damage is usually less rapid than in non-metals, although there have been reports of electron beam-induced migration of phosphorus during the STEM analysis of grain boundaries in steels. These 'smudging' effects can be all but eliminated if the probe is scanned along the boundary by aligning the scanning system with the

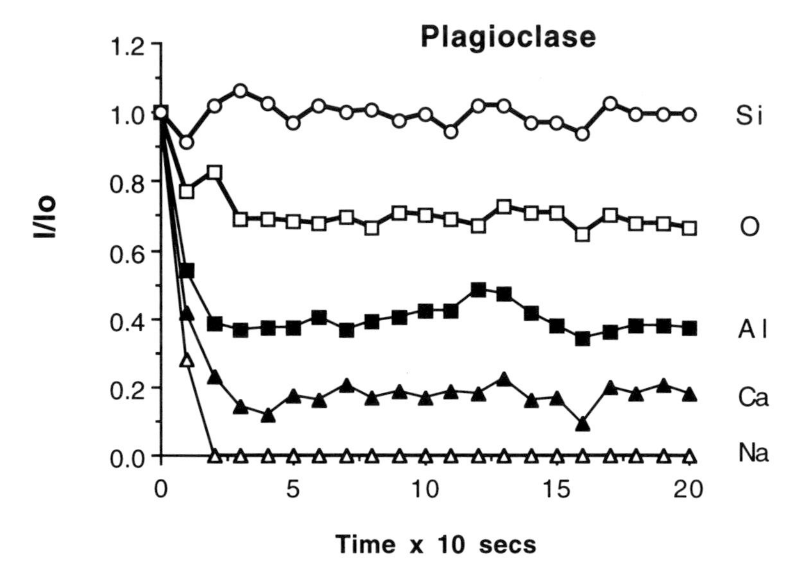

Figure 8.2. Relative loss of O, Na, Al, Si, and Ca from a 70 nm thick specimen of plagioclase feldspar irradiated at 100 keV with an electron current density of 1.28×10^6 A m^{-2}. (Courtesy of Dr P. E. Champness, University of Manchester.)

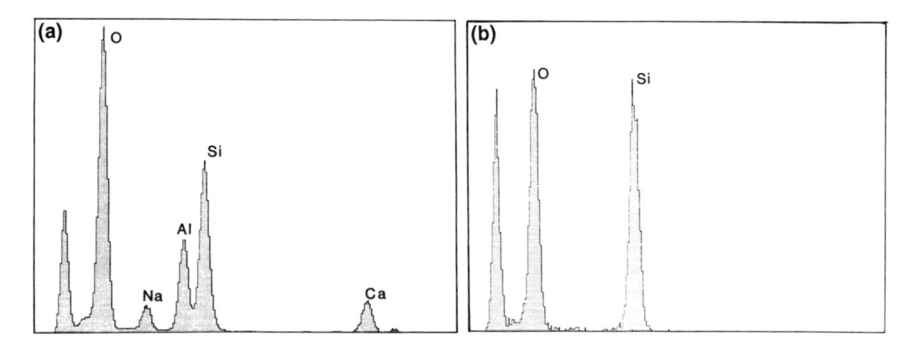

Figure 8.3. X-ray spectra from a plagioclase feldspar obtained (a) by rastering the defocused beam for 100 seconds across an area 2 μm square and (b) after exposure to a 100 kV focused probe at a current density of 1.8×10^8 A m^{-2}. (Courtesy of Dr P.E. Champness, University of Manchester.)

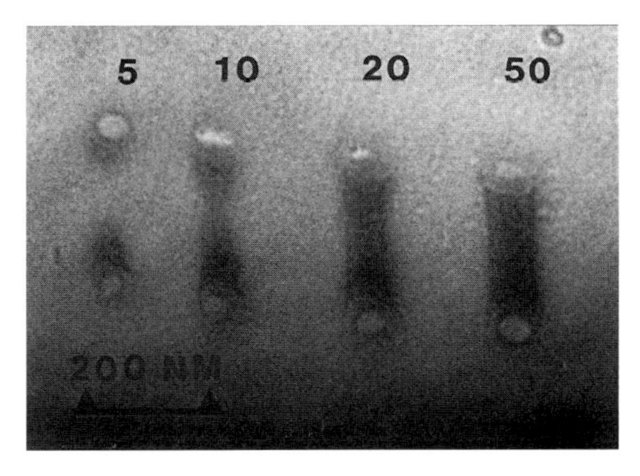

Figure 8.4. Damage columns in a plagioclase feldspar produced at 100 keV by a focused probe of current density ~5.6×10^5 A m^{-2} after exposure for 5, 10, 20, and 50 seconds. (Courtesy of Dr P.E. Champness, University of Manchester.)

boundary (using scan rotation). An alternative technique is to acquire an X-ray spectrum while the probe is scanned over an area which includes the boundary. If the probe is then scanned over an area of identical dimensions adjacent to the boundary for the same length of time, and this X-ray spectrum is subtracted from the first spectrum, the difference between the two will reveal any segregation at the grain boundary. Scanning an area also has the advantage that it allows an image of the boundary to be observed, so that specimen drift effects can be detected and counteracted.

While beam damage is a nuisance if one is attempting to obtain a high resolution image or an accurate chemical analysis, the damage can be used to carry out micro-machining on the nanometre scale. During extended irradiation modification of the composition is followed by loss of residual material from *both* the entrance and exit surfaces and a hole is drilled through the specimen, the diameter of which approximates the diameter of the focused probe. An example of the production of columns of damage in a feldspar is shown in *Figure 8.4*.

The scale of the features that can be 'manufactured' by using the FEG-STEM probe as a micro-machining tool are measured in nanometres, two orders of magnitude smaller than conventional sub-micron features produced on silicon chips using conventional technology. However, the time taken to carry out the micro-machining processes are far too long to contemplate using the technique yet in a manufacturing process.

Appendix A

Glossary

Some readers may have considerable experience of 'conventional' TEM but others may have little background knowledge of electron optics. It may help these readers if we define some of the terms used in the remainder of the book. Words shown in italics are defined elsewhere in the glossary.

Aberration: Any lens defect that results in the image of a point becoming distorted. *Spherical* and *chromatic* aberrations are present in every microscope and often limit its overall performance.

Angular magnification: Ratio between angle measured in image space and the corresponding angle measured in the object space. In our simple system (*Figure A1*) the angular magnification is given by u/v (i.e. the inverse of the lateral *magnification*).

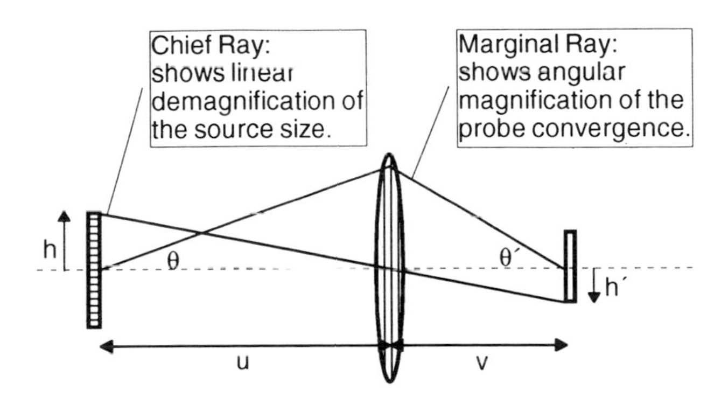

Figure A1. A simple ray diagram to illustrate relationships connected with probe forming and source size. Lateral magnification, m, is the ratio of h' to h (which can be shown to be the same as v/u). Angular magnification, M, is the ratio of θ' to θ (which, for small angles, can be shown to be the same as u/v). The well known lens formula can be applied to electron optical lenses, thus $1/u + 1/v = 1/f$.

Aperture: A metal (usually platinum or molybdenum) sheet containing one or more small circular holes used to stop all but the central chosen electrons from contributing to the beam. The holes are rarely more than 0.5 mm in diameter.

Astigmatism: A lens defect arising from asymmetry of the lens field which results in out-of-focus images appearing streaky and in-focus images appearing less sharp than they should. Astigmatism is corrected by a *stigmator* (either magnetic or electrostatic).

Beam diameter: Most beams of electrons are not sharp-edged ('top hat'), but their intensity varies across a diameter, possibly in a Gaussian profile. The beam diameter is usually defined to be that diameter which contains 50% (full width at half maximum for a Gaussian) of the total intensity in the beam. The central disc of a diffraction pattern generated by a circular aperture (Airy disc) contains 84% of the total intensity. Both distributions are shown in *Figure 4.2*.

Bremsstrahlung: From German meaning 'braking radiation', sometimes called 'white' radiation. When fast electrons come close to a nucleus the acceleration causes electromagnetic waves to be emitted. The energy of the photons exhibits a continuous distribution, but can reach the initial electron's kinetic energy.

Brightness: A measure used to compare the performance of electron sources. Often defined as the current density in a unit solid angle. Units: amps per square cm per steradian (A cm^{-2} sr^{-1}).

Camera length: In order to calculate an interplanar spacing, d, from diffraction pattern spot (or ring) radius, r, you require a camera length, L. Geometrically, $r/L = 2\theta_B = \lambda/d$; or simply $rd = \lambda L$, where λ is the electron wavelength and θ_B is the Bragg angle.

Chromatic aberration: Variation in lens focal length resulting from differences in the energy distribution of the waves being focused.

Coherent bremsstrahlung: Small 'peak' seen in the X-ray continuum background (*bremsstrahlung*), due to the interaction of the electrons with the periodic structure of a crystalline specimen, typically observed in the energy range 1–4 keV. It can be differentiated from a characteristic peak because the energy changes as the specimen is tilted, or the electron energy is varied.

Condenser: A *lens* or lenses used to control the convergence angle of a beam of electrons; also used to change system *demagnification*.

Cross-section: The probability of an event that may involve electrons changing their direction or velocity. Cross-sections are usually greatest for events involving the smallest change. Unit: square metres (m^2).

Demagnification: *Magnification* of less than unity (i.e. where the image is smaller than the object). *Condenser* lenses are often used to demagnify the electron beam before it hits the specimen.

Diffraction: Diffraction is simply the scattering of wave-like energy (e.g. light, X-rays, or electrons) from any non-uniformity of the medium (e.g. *aperture* edges, changes in specimen composition, individual (or collections of) atoms or molecules). Diffraction patterns from regular arrays of objects (e.g. atoms in a crystal lattice) are particularly easy and useful to interpret.

Electron lens: In virtually all electron microscopes electromagnetic lenses are used but their behaviour is described in terms of ray diagrams such as *Figure A1*. Magnetic lenses generally act as convex lenses.

Electron optics: The branch of optics that refers to the production and control of electron beams (cathode rays). Luckily most of the concepts of optics carry over to electron optics.

Field emission: The emission of electrons from a solid by quantum-mechanical tunnelling through the potential barrier of the work function. Field emission under high field gradients can lead to the emission of a large number of electrons even at room temperature. This process is the basis of the field emission gun (FEG).

Field emission tip: A very sharp single crystal of tungsten wire that is supported on a heatable hoop, so gases may be desorbed from the surface.

Focal length: The strength of a *lens* is usually defined in terms of its focal length, which is the distance from the lens at which a parallel beam is brought to focus.

Lens: A component of a microscope that produces a magnetic or an electrostatic field of changing strength through which electrons pass. The fields are controlled by external power supplies and shaped by elements such as magnetic pole pieces or electrodes.

Magnification: The magnification of an image is simply the linear size of a feature in the image divided by the linear size of the same feature in the object. The magnification of a lens is given by image distance over object distance (v/u, see *Figure A1*). (See *angular magnification*.)

Map: A two-dimensional representation of the composition of a specimen, obtained by recording the intensity of the counts in X-ray *windows* of different elements at different points in the specimen, and displayed as a grey-scale image.

Micrograph: A photograph taken using a microscope. A micrograph should always carry a scale marker to indicate the magnification.

Millibar: A unit of pressure, 10^{-3} bar. Atmospheric pressure is 1 bar, also equal to 100 000 Pa and 760 mmHg. One millibar is about 1.3 torr (where torr, named after Torricelli, is 1 mmHg).

Probe envelope: See *beam diameter*.

Projector lens: A lens used to magnify an image on to a screen. In CTEM the projector lenses control the image magnification and the diffraction camera length. In STEM the lenses are often referred to as post-specimen lenses and are most often used to control camera length and the electron-optical EELS interface.

Ray diagram: The ways in which lenses focus are illustrated by ray diagrams such as *Figure A1*. Such diagrams are usually two dimensional, which is satisfactory for the focusing of light. However, magnetic lenses also cause electrons to rotate around the optical axis. This effect is rarely shown in diagrams, and the same ray diagrams are used for light and electron lenses.

Region: A part of the *specimen* which can be seen in a single *micrograph*.
Sample: A (presumably typical) part of the whole object being studied, such as a piece of metal, or rat liver.
Specimen: A small piece of the *sample* chosen and prepared for study in the microscope.
Spectral image: As a map, but recording an entire spectrum, rather than a single elemental concentration at each pixel. Memory requirements are prodigious.
Spherical aberration: A lens defect that affects off-axis radiation by shortening the focal length of the lens with angle of incidence into the lens.
Stigmator: A part of the microscope which applies small quadrupole fields to correct *astigmatism* in a *lens* is called a stigmator.
Take-off angle: The angle between the surface of the specimen and a detector, typically the X-ray detector. Knowledge of the take-off angle is often required for quantitative analysis.
Thermionic emission: The emission of electrons from a solid as a result of heating. High temperatures are needed to give substantial emission and therefore materials such as tungsten are used. Thermionic emission forms the basis for the electron guns of most transmission and scanning microscopes. (See *field emission*.)
Ultimate resolution: The resolution of the instrument when all environmental and specimen related limitations are absent. Usually the theoretical resolution (where the diffraction and spherical aberration limits are equal) is taken as the ultimate resolution.
Ultra-high vacuum (UHV): A vacuum system capable of sustaining a specimen chamber vacuum level of better than 10^{-8} millibars over long periods of time, including specimen changes. Surface science applications require even better levels of evacuation than this.
Window: (i) Thin material covering the entrance to the X-ray detector, designed to provide vacuum isolation. Modern windows absorb very few of the incident X-rays, and are capable of withstanding atmospheric pressure. They are given the acronym 'ATM', standing for 'atmospheric thin window',
(ii) Small energy range selected from an X-ray or EELS spectrum, for the purpose of some form of numerical processing.

Appendix B

Further reading

Egerton, R.F. (1996) *Electron Energy-Loss Spectroscopy in a Transmission Electron Microscope*, 2nd Edn. Plenum Press, NY.
The most complete and authoritative account of the field.

Goodhew, P.J. and Humphreys, F.J. (1988) *Electron Microscopy and Analysis*, 2nd Edn. Taylor and Francis, London.
A sound introduction to electron microscopy.

Hren, J.J., Goldstein, J.I. and Joy, D.C. (eds) (1979) *Introduction to Analytical Electron Microscopy*, 2nd Edn. Plenum Press, NY.
Known as the 'green book', this excellent volume covers in greater detail the theoretical and practical basis for the techniques of AEM and includes sections on STEM by various authors.

Reed, S.J.B. (1993) *Electron Microprobe Analysis*, 2nd Edn. Cambridge University Press.
An excellent source for the fundamentals of EDXS.

Williams, D.B. and Carter, C.B. (1996) *Transmission Electron Microscopy*. Plenum Press, NY.
Widely acclaimed textbook based on a lecture course.

Other RMS handbooks (this series) that are particularly relevant:
3 *Specimen Preparation for TEM of materials*
20 *The Operation of Transmission and Scanning Electron Microscopes*

Computer programs

NIST Desk Top Spectrum Analyser 3.0 (reads all EDXS formats and models spectra from first principles, runs on a Macintosh).

GATAN EL/S 3.0 (program from world leading manufacturer of EELS systems, includes new cross-section database and several custom functions).

STEMSLICE (the program Kirkland at Cornell University wrote that simulates STEM images from an instrumental viewpoint).

Material Science on CD-ROM (paperback and CD-ROM), MATTER project (1996) ISBN 0-412-80080-2. Chapman and Hall, London. This software package contains several modules on microscopy and more will be added in the 1998 and future editions.

Journals to consult

Journal of Microscopy
Philosophical Magazine
Ultramicroscopy

Index

Aberrations, 4, 105
  chromatic, 40–41, 48, 51, 83
  gun, 14, 40
  objective, 18, 20 32, 40–41
  spherical, 39–41, 48–49, 56, 60, 67, 98, 99
Absorption (of X-rays), 28–29, 39–40, 51, 74, 76–77
Accelerating potential (voltage), 13, 14, 28, 29, 35, 39–40, 51, 78, 89, 98–99
Airlock, 33
Airy disc, 39, 106
Alignment
  condenser lens, 16–17
  gun lens, 11, 15
  objective lens, 18, 20
  post-specimen, 23, 60
  specimen (interface), 31–32, 102
Analysis, 3, 5–6, 9, 19, 27–30, 34, 75–77, 83, 85, 94, 97
Analytical electron microscope (AEM), 3, 11, 22, 26, 33, 71, 98, 109
Angular magnification, 40, 105
Annular dark field (ADF), 12, 23, 25, 42–44, 51, 59, 92–93
Aperture, 12, 16–17, 35, 37–39, 47, 51, 55, 58, 106
Astigmatism, 16–17, 19, 68, 106
Auger electrons, 17, 42, 46–47

Background, 22, 24, 28, 34, 38, 60, 73, 76–77, 86–87, 100
Backscattered electron, 5, 20, 23, 42, 50, 64
Bake-out, 33–34
Beam broadening, 34, 50, 78, 94
Beam damage, 8, 16, 26, 35, 49, 68, 89, 101ff
Beam diameter (*see also* Probe size), 6, 64, 105
Beryllium, 22, 71, 76, 79
Bragg angle, 25, 50–51, 53, 57, 59–60, 62–64, 66–67, 106
Bremsstrahlung, 28, 73, 76, 78, 86, 106
Bright field (BF)
  detector, 5, 12, 25, 42, 51, 56, 58–60, 65–67
  STEM image, 7, 24–25, 43–44, 50, 56, 66–67, 92, 96
Brightness, 1, 13, 15, 40, 48, 98, 106

Calibration, 31, 61, 76, 83, 85
Camera length, 60–62, 106
Cathodoluminescence, 42, 72, 80
Channels, 38, 72–74
Characteristic (radiation), 28, 42, 70–73, 76, 91, 98
Chromatic aberration, 14, 40–41, 48, 51, 83, 105–106
Cliff–Lorimer *k*-factors, 75–76, 100
Coherence
  imaging, 24–25, 43
  spatial (or lateral), 25, 66–67
Coherent bremsstrahlung, 74–75, 106
Collection angle, 23, 41, 43, 51, 60, 66, 96
Collector aperture, 2, 24–25, 42–44, 50, 59–61, 64–67, 81
Composition, 29–30, 41, 70, 76, 79, 85, 95–96, 101, 103, 106
Condenser aperture, 16, 21, 24, 43, 62
Condenser lens (C1 or C2), 2, 16–17, 21, 39–40, 49, 58–59, 106

Contamination, 16, 26, 33–34, 72, 78, 89, 93
Convergence angle, 11, 17, 21, 25, 39–40, 43, 48–50, 55–60, 62–64, 65–66
Convergent beam electron diffraction (CBED), 4, 24, 50, 62–66
Core (energy loss), 69, 85
Cross-over, 17, 19, 40, 59, 81, 83
Current density (*J*), 1, 13, 35, 39–40, 99, 101–103, 106

Dark field detector (*see also* Annular dark field), 5
Dead-time, 72, 78, 80
Defocus, 50, 56
Demagnification, 13, 17, 40, 48–49, 58, 98, 105–106
Detection sensitivity, 6, 25, 31, 72, 76, 89–90, 97–98
Differential pumping apertures (DPAs), 12, 15, 59
Diffraction, 3, 31, 39, 42, 55ff, 79, 81, 107
  contrast, 8, 21, 44–6, 51
  limit, 40–41, 47–49, 55, 66
  mode, 18, 21, 58–59
  pattern, 4, 21, 24, 52, 55–58, 60–61, 62–65, 81, 106
  pattern observation screen (DPOS), 12, 24–25, 42, 50, 60, 65
Dwell time, 41, 52–53, 92

Elastic scattering, 50
Electric field, 12, 22, 39
Electron dose (*see also* Current density), 8
Electron energy loss spectroscopy (EELS), 1, 5, 9, 23, 25–27, 31, 51, 61, 64, 69, 80ff, 99
  fine structure, 84, 87–88
  PEELS, 79, 83, 85, 98
  spectrometer, 25–27, 82–83
  spectrum, 81, 84–85, 87–88
Electrons (as waves), 25, 38, 47, 62, 68, 81, 99, 106
Electrostatic
  focusing, 13–15, 107
  scanning, 18
  stigmator, 19
EMMA, 80, 97–98
Energy dispersive X-ray spectroscopy (EDXS), 2, 6, 25, 27–30, 69ff, 89–90,101
  analysis, 8, 27, 79, 90
  linescans, 6, 45, 94, 101
  mapping, 18–19, 24, 4290, 91ff, 93, 100
  spectrum, 7, 51, 69–70, 72
Energy filtering, 25, 68, 82, 84
Energy level (or state), 25, 69, 84
Energy loss near edge structure (ELNES), 87, 101
Energy resolution, 25, 31, 72–73, 79, 84
Escape peak, 74
Extended energy loss fine structure (EXELFS), 88
Extraction electrode, 14

Far field, 22, 59–60
Field emission (FE), 1, 26, 107
  FEG-STEM, 6, 11, 35, 51, 78, 98, 101–102
  gun (FEG), 12–14, 34–35, 40, 67, 82, 89, 99
  tip, 13–14
Focal plane, 21, 51–52, 55–59, 60, 82
Fourier methods, 88–89
Full-width at half-maximum (FWHM), 13, 48, 50, 95, 106

Gaussian, 14, 39, 49, 73, 78, 95, 106
Geometry, 5, 27–30, 42, 51, 61, 74, 76
Grigson coils, 23, 64
Gun lens, 14–15, 40–41, 59

High-angle annular dark field
  detector (HAADF), 24, 42, 59–60
  imaging (HAADFI), 45, 50, 66, 91, 95ff
High voltage, 15, 28
Hole-count, 19

Image, 20, 37ff, 41, 53, 56–57, 78, 91
  coherent, 24–25, 43
  incoherent, 46
  mode, 17–18, 65–66
  plane, 52, 55
Inelastic scattering, 89
Interference
  coherent, 66–67, 68
  external, 18, 38–39, 48, 48, 72

Ionization, 69, 84, 86

Kikuchi lines, 60, 62

Lateral
  coherence, 25
  magnification, 105
  movement of stage, 22
  spreading of beam, 34, 50
Lattice image, 19, 21, 45, 50, 62, 65–67
Lens, 78, 99
  aberrations, 39–41, 68
  bore, 19, 22–23, 42
  condenser, 16–17, 39–40, 49, 58–59, 106
  electrostatic, 15
  gap, 14, 20, 22–23, 32
  gun, 14–15, 40–41, 59
  objective, 19–23, 31–32, 39–40, 42, 48, 52, 55, 57–59, 61–62, 82
  post-specimen, 4–5, 23, 41–42, 51, 55–57
  projector, 2, 23, 57–58, 82
Light element analysis (*see also* Absorption), 6, 28, 34, 71–72, 79
Linescan, 45, 101

Macromolecules, 8–9
Magnetic
  field, 14, 18, 20, 22–23, 39, 81, 83, 107
  lens (*see* Lens)
  sector, 81
  shielding, 18, 26
Magnification, 3, 18–19, 24, 31, 38, 57–58, 61, 65–66, 92, 107
Mapping (*see* X-ray mapping)
Microanalysis (*see also* Analysis), 4, 27, 48, 69ff, 78–79, 99
Microdiffraction, 62–65
Multiple scattering, 51, 89, 101

Objective aperture (real), 2, 17, 21, 42, 49, 51, 56, 59–62, 65
Objective lens (OL), 2, 17, 20, 22, 39–40, 48, 52, 55, 57–59, 61, 66

Parallel electron energy loss spectrometer (PEELS, *see also* EELS), 2, 79–90, 98, 101
Penetration, 51, 99
Phase contrast, 43
Photomultiplier tube (PMT), 2, 24–25, 41
Pixel, 4, 38, 41, 53, 92, 101, 108
Plasmon, 85–86
Point spread function, 47
Pole-pieces, 20, 22, 82, 107
Post-field (objective), 20, 22–23, 53, 59–62, 65
Post-specimen
  alignment, 23
  compression, 20, 61, 83
  lenses, 4–5, 23–24, 41–42, 51, 55–56, 62
Potential barrier, 14
Pre-field (objective), 17, 22, 65
Probe envelope, 39, 42, 50, 56
Probe size (probe diameter), 1, 11, 13, 27, 34, 38–40, 47–50, 55, 62, 66–67, 70, 78, 89, 96–99
Process time, 80
Projector lens, 2, 23, 57–58, 83, 107
Pulses, 70, 80, 91

Quadrupole lenses, 20
Quanta, 69, 84
Quantitative analysis, 30, 75–77, 83, 85, 101
Quantum tunnelling, 13–14

Raster pattern, 18, 24, 37
Ray diagrams, 17, 40, 52, 59, 65, 107
Rayleigh criterion, 47
Reciprocity, 4, 56
Resolution, 4, 6, 24, 27, 32, 39–42, 47
  diffraction, 55, 57, 61, 64–65, 68
  energy, 25, 31, 72, 79–80, 82
  image, 25, 47, 50–51, 56, 66–67, 96, 103
  microanalysis (spatial), 50, 75, 78–79, 97
  ultimate, 4, 47–48, 55, 68, 108
Rocking point, 17, 21, 59–60
Rutherford scattering, 24

Scan coils, 11, 17–18, 21, 59, 65
Scanning electron microscope (SEM), 1, 11, 22–23, 37, 57–58, 70, 80, 100
Scanning modes
  diffraction, 18, 21, 58–59

image, 18, 58–60, 103
spot, 60, 65
Scintillator, 22, 24–25, 41, 51, 59–60
Secondary electron detector (SED), 1, 5, 22, 42
imaging, 44, 46
Selected area diffraction (SAD), 58–61, 65
Selected area diffraction aperture (SAD ap.), 16–17, 19, 58
condenser astigmatism, 16
hole-count, 19
position, 2, 12
ray diagram, 17, 59
rocking point, 21, 59
Slit (EELS), 81–82, 83–84
Solid angle, 27, 29
Source size, 13–14, 40, 48–49, 98, 105
Spatial resolution (*see also* Probe size), 6, 27, 34, 78, 97
Specimen, 2, 19, 28–29, 56–58, 69, 76–78, 91, 108
density, 70, 76
drift, 18, 22, 31, 33, 92,103
holder, 11, 20–22, 61
thickness, 27, 29–30, 34, 50, 68, 70, 76–77, 80, 89, 92, 94
Spectral imaging (*see also* X-ray mapping), 100ff, 108
Spherical aberration (*see also* Aberrations), 39–41, 98, 99ff, 105, 108
Stage
side-entry rod, 2, 21, 26, 33
top-entry cartridge, 2, 21, 23
Statistics, 78
Stigmator, 16, 19, 106, 108
Sum peak (X-ray), 74–75

Take-off angle, 28–29, 76, 108
Thermionic, 13–15, 80, 82, 89, 98–99, 108
Tilt (of specimen), 22, 29–32, 40, 61, 83, 106
Tip (FE), 12–14, 48, 107
Transmission electron microscope (TEM), 1, 3–5, 11, 19, 21, 24–26, 27, 41, 43, 50–52, 55–58, 60–62, 65–66, 68, 70, 83, 96–97, 100, 109

Ultra-high vacuum (UHV), 3, 15, 108

Vacancy, 69
Vacuum, 14, 33, 35, 70–72, 108
Virtual objective aperture (VOA), 2, 15–19, 39–40
convergence angle, 17–18, 40, 48–49
hole-count, 19
probe defining, 15–16, 39, 48–49
real objective, 21, 60

Wavelength (X-rays), 79, 97–98
Window, 71, 79, 91–92, 108
Windowless X-ray detector, 33, 72–74
Wobbler, 20
Work function, 13–14, 99, 107

X-ray
detector, 1, 5, 22–23, 27–29, 31, 70–71, 75, 97–98, 108
linescan, 45, 94
mapping, 18–19, 24, 42, 90, 91ff, 93, 100

Yttrium aluminium garnet (YAG), 41

Z-contrast (atomic number contrast), 50, 96
z-shift, 22, 31–32, 83
Zero-energy peak (EDXS), 73
Zero-loss peak (EELS), 83–86